T0240771

Lecture Notes in Computer Science   8820

Commenced Publication in 1973
Founding and Former Series Editors:
Gerhard Goos, Juris Hartmanis, and Jan van Leeuwen

More information about this series at http://www.springer.com/series/7409

Giulio Jacucci · Luciano Gamberini
Jonathan Freeman · Anna Spagnolli (Eds.)

Symbiotic Interaction

Third International Workshop, Symbiotic 2014
Helsinki, Finland, October 30–31, 2014
Proceedings

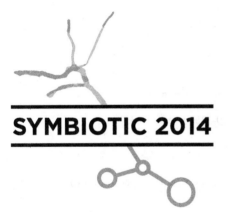

 Springer

Editors
Giulio Jacucci
Department of Computer Science
Helsinki Institute for Information
   Technology HIT
University of Helsinki
Helsinki
Finland

Luciano Gamberini
University of Padua
Padua
Italy

Jonathan Freeman
Department of Psychology
Goldsmiths, University of London
London
UK

Anna Spagnolli
University of Padua
Padua
Italy

ISSN 0302-9743   ISSN 1611-3349   (electronic)
Lecture Notes in Computer Science
ISBN 978-3-319-13499-4   ISBN 978-3-319-13500-7   (eBook)
DOI 10.1007/978-3-319-13500-7

Library of Congress Control Number: 2014956516

LNCS Sublibrary: SL3 – Information Systems and Applications, incl. Internet/Web, and HCI

Springer Cham Heidelberg New York Dordrecht London

Printed on acid-free paper

Springer International Publishing AG Switzerland is part of Springer Science+Business Media
(www.springer.com)

Preface

Symbiotic 2014 is the third international workshop on symbiotic interaction. The first workshop was held in Padua during December 3–4, 2012. It was chaired by Prof. Luciano Gamberini of the University of Padua. The workshop provided interesting dialog with industry members, including, start-ups, small, and large companies. The second workshop was held on December 12, 2013 and brought together academics working on ubiquitous computing and brain computer interfaces. The workshop was chaired by Prof. Jonathan Freeman of Goldsmiths, University of London.

This third edition of Symbiotic 2014 marks a change in the series, as the first-time proceedings of the workshop are published. We solicited 16 high-quality submissions in three categories: papers, posters, and demos. The workshop gathered a long list of important scholars in many disciplines (see Program Committee), and each anonymous paper was reviewed by three members. We accepted eight full papers, three posters, and two demos.

The first paper in the proceedings is an introduction by the Chairs and Program Chairs of the concept of Symbiotic Interaction including an overview of submissions.

We believe that Symbiotic will continue to grow and attract more interest from disparate fields with the aim of investigating future relationships between computers and humans. Symbiotic 2014 is partly funded by the MindSee Project and is partially funded by the European Community (FP7 – ICT; Grant Agreement # 611570) and by TEKES, the Finnish Funding Agency for Innovation through the Re:Know Project.

October 2014

Giulio Jacucci
Luciano Gamberini
Jonathan Freeman
Anna Spagnolli

Organization

General Chairs

Giulio Jacucci   University of Helsinki, Finland
Luciano Gamberini   University of Padua, Italy

Program Chairs

Jonathan Freeman   Goldsmiths, University of London, UK
Anna Spagnolli   University of Padua, Italy

Program Committee

Dan Afergan	Tufts University, USA
Elisabeth André	University of Augsburg, Germany
Rosa María Baños Rivera	Universitat de València, Spain
Joydeep Bhattachyara	Goldsmiths, University of London, UK
Frank Biocca	Syracuse University, USA
Benjamin Blankertz	Technische Universität Berlin, Germany
Marc Cavazza	Teesside University, UK
Manuel Eugster	Aalto University, Finland
Marco Filetti	Aalto University, Finland
Jodi Forlizzi	Carnegie Mellon University, USA
Jonathan Freeman	Goldsmiths, University of London, UK
Luciano Gamberini	University of Padua, Italy
Hans Gellersen	Lancaster University, UK
Dorota Glowacka	University of Helsinki, Finland
Christoph Guger	g.tec Medical Engineering, Austria
Sumi Helal	University of Florida, USA
Wijnand IJsselsteijn	Eindhoven University of Technology, The Netherlands
Rob Jacob	Tufts University, USA
Giulio Jacucci	University of Helsinki, Finland
Samuel Kaski	Aalto University, Finland
David Kirsh	UC San Diego, USA
Jane Lessiter	Goldsmiths, University of London, UK
Rod McCall	Université du Luxembourg, Luxembourg
James Moore	Goldsmiths, University of London, UK
Roderick Murray-Smith	University of Glasgow, UK
Niklas Ravaja	Aalto University, Finland
Andreas Riener	Johannes Kepler University Linz, Austria

Tuukka Ruotsalo	Aalto University, Finland
Erin Solovey	Drexel University, USA
Anna Spagnolli	University of Padua, Italy
Michiel Sovijärvi-Spapé	Aalto University, Finland
Oliviero Stock	IRST – Istituto per la Ricerca Scientifica e Tecnologica, Italy
Paul Verschure	Universitat Pompeu Fabra, Spain
Zhao Xuran	Aalto University, Finland

Web and Communication Chairs

| Oswald Barral | University of Helsinki, Finland |
| Khalil Klouche | University of Helsinki, Finland |

Keynotes

Contextual Robotics

David Kirsh

University of California, San Diego

Abstract. Robots will soon be among us in diverse forms. They will be expected to blend in. Humans take for granted our remarkable sensitivity to situational factors. Robots will have to learn these. When we act together we respond to social cues that are both explicit – speech, gesture, nodding, pointing – and implicit – eye movement, gaze, where, when and how someone stands, oves, makes eye contact. When we cooperate on a task we need to be aware of each other's goals and objectives. We need to coordinate activity. In this talk I will look at some of the diverse ways people coordinate activity when they act jointly and talk about how detection and response to these cues represents another hurdle for new age robots.

David Kirsh is Professor and past chair of the Department of Cognitive Science at UCSD. He was educated at Oxford University (D.Phil), did post doctoral research at MIT in the Artificial Intelligence Lab, and has held research or visiting professor positions at MIT and Stanford University. He has written extensively on situated cognition and especially on how the environment can be shaped to simplify and extend cognition, including how we intelligently use space, and how we use external representations as an interactive tool for thought. He runs the Interactive Cognition Lab at UCSD where the focus is on the way humans are closely coupled to the outside world, and how human environments have been adapted to enable us to cope with the complexity of everyday life. Some recent projects focus on ways humans use their bodies as things to think with, specifically in dance making and choreographic cognition. He is co-Director of the Arthur C. Clarke Center for Human Imagination, and he is on the board of directors for the Academy of Neuroscience for Architecture.

Models and Measures of Human–Computer Symbiosis

Roderick Murray-Smith

University of Glasgow

Abstract. This talk will be a response to the review paper of Jacucci et al.. I will try to frame Human-Computer Symbiosis in an abstract model, and then examine the role of control theory and information theory in measuring the level of symbiosis in any given human-computer dyad. I will also explore the importance of the symbiotic stance to understanding contextual niches and ecosystems of services involving multiple users and multiple services.

Roderick Murray-Smith is a Professor of Computing Science at Glasgow University, in the "Inference, Dynamics and Interaction" research group, and is the Director of SICSA, the Scottish Informatics and Computing Science Alliance. He works in the overlap between machine learning, interaction design and control theory. In recent years his research has included multimodal sensor-based interaction with mobile devices, mobile spatial interaction, Brain-Computer interaction and nonparametric machine learning. Prior to this he held positions at the Hamilton Institute, NUIM,Technical University of Denmark, M.I.T., and Daimler-Benz Research, Berlin. He works closely with the mobile phone industry, having worked together with Nokia, Samsung, FT/Orange, Bang & Olufsen and Microsoft. He is a member of Nokia's Scientific Advisory Board. He has co-authored three edited volumes, 19 journal papers, 16 book chapters, and 76 conference papers.

Contents

Demos and Posters

Definitions of Symbiotic Interaction

Symbiotic Interaction:
A Critical Definition and Comparison
to other Human-Computer Paradigms

Giulio Jacucci[1,3(✉)], Anna Spagnolli[2],
Jonathan Freeman[4], and Luciano Gamberini[2]

[1] Department of Computer Science, Helsinki Institute for Information
Technology HIIT, University of Helsinki, Helsinki, Finland
giulio.jacucci@helsinki.fi
[2] Department of General Psychology, University of Padua, Padua, Italy
{anna.spagnolli,luciano.gamberini}@unipd.it
[3] Helsinki Institute for Information Technology HIIT,
Aalto University, Espoo, Finland
giulio.jacucci@hiit.fi
[4] Department of Psychology, I2 Media Research,
Goldsmiths University of London, London, UK
J.Freeman@gold.ac.uk

Abstract. We propose a definition of symbiotic interaction that is
informed by current developments in computing. We clearly distinguish
this definition from previous ones and from selected paradigms that
address the human-computer relationship. The definition is also informed
by a variety of human-centered frameworks in human-computer interaction, including embodied interactions, situationist frameworks, and
participatory and work-oriented design perspectives. Symbiotic interactions can be achieved by combining computation, sensing technology,
and interaction design to realize deep perception, awareness, and understanding between humans and computers. Important aspects to implement are transparency, reciprocity, and collaborative use of resources for
both computers and humans. The symbiotic relationship is also characterized by goals and agency independence of humans and computers.
The definition sets the premise to discuss in a critical way future research
agendas for symbiotic interactions that are sensitive to human-centered
values.

Keywords: Human-computer symbiosis · Human-computer interaction
frameworks · Research agenda · Affective computing · Persuasive
technologies · Embodied interaction

1 Introduction

Symbiotic interaction identifies interactive computerized systems that create
steps further with respect to user-centered design paradigms. Symbiotic interaction relies on a new generation of resources to understand users and to make

© Springer International Publishing Switzerland 2014
G. Jacucci et al. (Eds.): Symbiotic 2014, LNCS 8820, pp. 3–20, 2014.
DOI: 10.1007/978-3-319-13500-7_1

themselves understandable to users. We refer to the fact that distributed systems can easily and autonomously sense the physiological and behavioral information of users over time; it can aggregate and analyze this and other information in a large mass of users. This makes it possible to design systems that detect the users' psychophysiological states. This also makes it possible to predict actions that use this information to better adapt output to the user regardless of his/her ability to explicitly refine his/her request. In addition to relying on these technical developments, symbiotic systems can benefit from a set of paradigms and reflections that were developed over the last 3 decades to define how and why human actions and experiences are affected by computers. In this introduction, we propose a definition of symbiotic interaction that although indebted to J.C.R. Licklider's phrase, human-computer symbiosis, which appeared in 1960 aims to capture current developments in computing [1]. We start by distinguishing this definition from previous ones that use the symbiosis term and that are from selected paradigms that address the human-computer relationship, such as persuasive technologies, affective computing, and telepresence. To ensure a critical and human value sensitive approach, the definition is also informed by a variety of human centered frameworks in human-computer interaction. This includes embodied interactions and situationist frameworks, as well as participatory and work-oriented design perspectives. In the discussion, we reflect in a critical way on future research agendas for symbiotic interaction that are sensitive to human-centered values. We also consider challenges and opportunities and pinpoint a specific set of ethical concerns.

2   Previous Use of Symbiosis in Human-Computer Paradigms

The concept of man-computer symbiosis [sic] dates back to 1960 when Licklider [1] published an article that created an analogy to biology, in which an insect and fig tree needed each other in a productive and thriving partnership. The aim was to investigate a vision in which human brains and computing machines would be coupled very tightly. The vision, Licklider argued, is different from extended human (or what later would have been known as augmented human by Engelbart [2]) as the human dominates as a sole organism. Also, it is different from artificial intelligence, where the computer dominates as the principal problem solver. Licklider analyzes capabilities of humans and machines and envisions which functions of human and computers are or are not separable in symbiotic association, starting from assigning mechanical retrieval tasks to computers to providing free time for humans for actual problem solving or decision making. Humans will set goals, motivations, and formulate hypotheses. To the contrary, computers convert hypotheses into testable models, models against data, transform data, perform statistical inferences, and work as decision theory or game theory machines. This vision was mostly guided by information and cognitive processing perspectives. The vision is, to some extent, realized in a variety of software tools that are currently used in desktop and mobile computing. In 2004,

Roy proposed a program at MIT with the goal of magnifying human abilities by order of magnitude (10) under the concept of human-machine symbiosis [3]. Here, the most important aspects of the research agenda were perceptual computing and the capability of systems to sense the world; natural embodiments and representations of human-friendly machines; and learning systems beyond instructions and programming. Human performance would then be magnified by memory, expression (translation of intention into action), listening, (selectively attending to a variety of inputs), learning, understanding, physical skills, and awareness. Erp et al. [4] directly apply the concept of user system symbiosis to brain-based interfaces and EEG. They use a model based on perceptual control theory that provides insight into what happens in high workload situations and how this affects physiological and brain-based indices, suggesting that a symbiotic system should intervene in overload situations. Schalk [5] introduces the concept of brain-computer symbiosis to overcome the limitations of our bodies' input and output capacities by using direct interactions with the brain. This view is guided by the speed of communication and the analyses of different interfaces and their communicative powers. Both decoding information from the brain and inducing information into the brain are considered. The idea of symbiotic systems that we would like to capture unifies these inputs into a comprehensive category that does not refer to any specific interface or application. We recognize that humans' and computers' capabilities work in sole unity. Human performance is augmented without losing control, and this attributes to the computers' ability to implicitly (even subliminally [6,7]) detect the users' states and goals, which are the main differences with prior user-centered paradigms.

3   Relevant Frameworks on Human-Computer Interdependence

Several frameworks focus on human-computer interactions or system designs and emphasize the interdependence between humans and machines in a way that is relevant to the symbiotic paradigm outlined here. In this section, we will mention 4 of them, which are instrumental to highlighting the principles that, in our opinion, characterize the mediated experience in symbiotic systems: (a) telepresence, as the study of co-constructing between humans and machines in the experience of being in a mediated space; (b) mixed initiative interaction, as a framework to predict and act on users' intentions; (c) affective and emotional computing, as the capability of computing systems to recognize or affect emotions [8]; and (d) persuasive technology, as the embedded ability of computers to direct human behavior [9].

3.1   Telepresence

Presence or telepresence is a notion used in human-computer interaction and media studies. This consists of the subjective experience of being in a mediated environment [10,11], namely an environment that is mainly supported by digital resources.

Glossed as the sense of being there, this experience is described with the metaphor of transporting the user into the mediated world or, vice versa, of transporting digital objects and characters into the user's environment [12]. The experience of presence incorporates at least 2 different kinds of phenomena: the consistent feeling of being in a specific spatial context and intuitively knowing where one is with respect to the immediate surrounding (spatial presence [13]); and the sense of being with another person while mediated by telecommunication technology (social presence, [14]). Presence has been conceptually conceived according to different and often competing models. Spagnolli et al. [15] highlight that most presence measures imply that presence is either an illusion or an attentional phenomenon; in both cases, separation between real and digital is assumed, and the former is attributed to a stronger ontological status. Adopting either of these approaches leads to the classic division of body and mind. This is a view that has been criticized because it only fits forms of mediated presence that are immersive. It has been criticized also because it does not account for the physical resources that are necessary parts of the experience during telepresence, involving, for instance, the body or an input device [16]. Constructionist, phenomenological, or pragmatist concepts of presence (e.g., [17,18]) are instead closer to the notion of symbiosis, since they show that being present in a digital, mediated environment is necessarily a hybrid experience; it results from intertwining digital and non-digital resources that are interdependent from each other in the constitution of the mediated experience [16].

Reflection. (Tele)Presence is one concept that—along with other concepts, such as hybrid or cyborg—conveys the intimacy of the relation between humans and the tools that they use. As such, the concept captures one core aspect of the notion of symbiosis, namely the interdependence between the units in a symbiotic relation. Focusing on presence, while designing a system, forces one to find strategies to intensify this symbiosis. Some of the strategies are particularly in line with the idea of a symbiotic system (i.e., getting an immediate and coherent response to input that changes the environment itself) other less (i.e., realism). Instead, presence does not overlap with symbiosis to the extent to which the latter regards the human experience and does not directly contribute to designing the relationship between humans and machines in a dynamic way, in which machines adapt to the users.

3.2 Affective Computing

Affective computing refers to computing that relates to, arises from, or deliberately influences emotions [8]. Generally, this translates into research on affective input and affective output of a computing system. For the affective input, recently, work has concentrated on multimodal fusions of different modalities. Recently input processing has extended to physiological measures, including brain signals. In terms of using physiological sensors to adapt an information environment, these have been based previously, for example, on modeling affective states to adapt narratives or visualization features in interactive art.

In Gilroy et al. [19] affective fusions of GSR, EMG, and speech are used to model affective states in a computational dimensional model of effects, including pleasure, arousal, and dominance. The adaptation results in size and coloring of interactive 3D art. Another stream of research in affective computing is the development of embodied agents that are capable of emotional expressions, as an important step is to reproduce affective and conversational behaviors [20,21].

Reflection. Affective computing is relevant to symbiotic interaction, as it concentrates on detecting or affecting emotions, which can deepen the symbiotic relationship between humans and machines. However, this is a partial view of a symbiotic relationship, since the affective component is seen as just one of the aspects that a symbiotic system should consider.

3.3 Persuasive Technologies

Computer applications and devices can support people in changing their habits. They can direct users towards specific behavioral options in a variety of application areas: sustainable choices and pro-environmental awareness [22,23], health and wellness [24,25], safety [26]. These applications exploit psychological theories of persuasion and have been recently identified as a unified domain due to the work of B.J. Fogg, who coined the term persuasive technology. The work of B.J. Fogg [9] and the growing community around it are of particular interest to a symbiotic paradigm for at least two aspects. The first is the remark that a machine has an embedded, implicit persuasive power that rests on its very design. This remark is based on the older notion of affordance, to which the features of a technical tool invite a certain action on the user's part; while the notion of affordance has mainly inspired efforts in the area of usability, exploiting affordances for persuasive purposes is the next step. The second aspect of persuasive technologies—that is of interest to a symbiotic framework—is the ethical concern for the user's ability to actually control his or her choices vis-a-vis the persuasive power of technical tools [27–29]. In the field of persuasive technology, the solution has been to exclude coercion or deception from the domain by definition and then rule out possible misusages. In addition, the dominant models of persuasive technology have been criticized for being too focused on shaping human behavior deterministically, which is both an ethical issue and a risk for the success of the persuasive effort [30]. Recently, researchers have criticized these models [31] and have argued for other paradigms that more explicitly state how computers should empower users' choices, instead of prescribing actions.

Reflection. Symbiotic systems have a persuasive component in that they direct the users' choices when implicitly reconfiguring their output. They prioritize options based on an understanding of the users' goals and preferences. Although this is meant to increase performance, while preserving the users' priorities in the mediated activity, it surely presents ethical issues related to users' agency and to the transparency of the systems' criteria.

3.4 Other Frameworks

A notable related approach is mixed initiative interaction by Horovitz [32]. It provides an alternative way to predict a user's intentions. It refers to a flexible interaction strategy, wherein an agent can contribute resolving the user's task in an interactive manner. It can initiate a dialogue for the user when it infers that the user may need assistance in navigation or problem solving. In mixed-initiative interaction, both the user and the system are allowed and expected to be active. However, it has been traditionally developed as part of an agent system to assist users in an office tools environment. In this environment, its success has been limited by the efforts required from the user to correct prohibitive inferences and dialogue propositions mistakenly initiated by the system. Besides symbiosis or augmentation, other metaphors have been explored, such as the H-metaphor for vehicles, automated systems, and robotics that uses horse riding as a way to explain optical human-machine relations [33]. In principle, a variety of other movements, frameworks, and communities could be considered, but they are not conceptual paradigms. Exemplarily adapted interaction [34] includes conferences (UMAP Conference on User Modelling, Adaptation and Personalization) and journals (UMUAI or user modeling and user-adapted interaction) concerned with user adaptive systems and user modeling that are relevant for symbiotic interaction research agendas.

4   Critical Contribution from Human-Centred Frameworks

4.1 Skills and Resourcefullness

Traditional human-computer interaction approaches to evaluate the usability of products or computers for people tend to view the person as a user and the computer as a tool, and the latter is used to accomplish a task (cfr. [35]). Before the attention to experience design [36] HCI research has been infused with frameworks that investigate the situated character of our actions, how plans are used as resources [37], distributed characters of cognition and actions [38], the historical aspects of practice, and the mediating functions of artefacts [39]. Carrol and Rosson [40] propose the concept of active users, which highlights the view that users cannot be represented as information-processing automata that merely generate responses to stimuli provided by an interface. Other work reflects the situated character of our actions and how plans are used as resources, particularly considering how these are inscribed or integrated in a user interface [37]. Phenomenological approaches have inspired different perspectives of actions related to technology use as the notions of involved unreflected activity and breakdown [41]. Ehn [42] presents a different explanation of practices of design and use, while using the language games approach of Wittgenstein and the notion of family resemblance. Ehn also discusses the consequences of considering computer artefacts as tools. Artefacts are objects that are made by human work. In designing computer artefacts, the emphasis should be on concernful designs of signs that make sense in the language game of use (p. 164). Computer artefacts should be seen as mediating instrumental and/or communicative activities; as supporting individual and/or cooperative activities; as augmenting and/or replacing human activities; and as functions and forms that are irrevocably interconnected (p. 165) According to Ehn, computers should not be considered just as tools but as designed artifacts that recognize the importance of skills.

Implication. An important implication is the notion of skill and how a technological actor, such as a computer, should enhance and develop skills, instead of just automating actions.

4.2 Manifestation and Accountability: Embodied Interaction

Dourish presents a view on how computation should be manifested through the use of explicit, causally-connected self-representations in computational systems. These are embodied through accounts in which systems offer their own activities with consequences for the system designs and interactions in supporting resourceful actions [43]. In embodied interactions, the active nature of computers is important not as an independent agent but as an augmentation and amplification of our own activities [44]. Dourish [45] stresses that embodied approaches to interactive systems allow us to: engage with technology in different ways that allow us to uncover, explore, and develop the meaning of the use of technology as it is incorporated in practice. In turn, this suggests that the meaning of the use of the technology is, first, in flux and second, is something that is worked out again and again in each setting. If this is true, then the technology needs to be able to support this sort of repurposing and needs to be able to support the communication of meaning through it (p. 239). Central to Dourish's proposition is the notion of manifestation. Dourish presents a view on how computation should be manifested through the use of explicit, causally-connected self-representations in computational systems. These are embodied through accounts in which systems offer their own activities with consequences for both the system designs and interactions in supporting resourceful actions [43].

Implication. A key input is the opportunity to consider how computation should be manifested in terms of providing support for resourceful interactions of users. Additionally, transparency and accountability have important roles in this view.

4.3 Design and Computers as Theater

Several design-oriented frameworks have been proposed as the one of Redström [46], who proposes a design philosophy for everyday computational factors, where meaningful presence is contrasted to previous imperatives from usability as, for example, efficient use. In this design approach, time is the central parameter, as exemplified by slow technology [47], and aesthetics is the basis to design presence. Redström describes the presence of an artefact in terms of how it expresses itself as we encounter it in our everyday life. Then, we can think of artefacts as expressions, artefacts, and bearers of expressions, rather than functions [47]. Phenomenological approaches have also inspired design frameworks [48]. All of these contributions, as well as attention to performative approaches as design techniques in creating and evaluating scenarios [49–51] created the premise to explore performances and theatrical frameworks in HCI. Laurel's work [53] was the first to consider theatrical metaphors and drama in the context of human-computer interactions by applying the principles of Aristotelian poetics. Human-computer experiences can be structured around the precepts of dramatic forms and structures. Laurels' aim is to derive poetics from interactive forms. Here, interactivity is understood as the ability of humans to participate in actions in a representational context.

Implications. As implications we conclude that besides tangible, expressionals, perceptive, and slow technology design frameworks, recent interaction design movements include performative perspectives; these address the roles of users as actors and audiences, the dramaturgy of trajectories of experience, and how fictional spaces emerge in interactions between users and interactive systems [52, 54–56]. All of these aspects are important to consider in developing symbiotic interaction systems.

4.4 Ethics

Symbiotic systems rely on technology that might be able to acquire and elaborate data from a large mass of users; this information might be sensed without users being aware of it and without them being able to control its provisions; finally, these systems have a model of the users based on prior behaviors and provide cues that direct the users' choices in order to improve their performances. Although these features are necessary to deliver the kind of services promised by symbiotic systems, they also raise ethical issues because of their potential threat to privacy and the possible misusage of the processed data. These risks are enhanced when the information gathered is not only personal (i.e., relating to or traceable to an individual person) but also sensitive. Namely, this refers to data, such as medical information, that might result in loss of advantages or level of security if disclosed to others who might have low or unknown trustability or undesirable intentions. Symbiotic systems share these risks with other information technology but also magnify the risks connected to the users' loss of control and agency. This is because they rely on information that is captured implicitly and because they pre-filter the information to be shown to users. These risks can be summarized as follows:

- User's identification based on collected data [57].
- Permanence of personal/sensitive data after research or project Conclusion [58].
- Profiling (by merging databases, by data-mining) and attribution of new property to the individual derived from implicit pattern [59–61].
- Use of data for monitoring [62] and surveilling sanctionable acts [63].
- Misinterpretation of data disregarding their original meaning when produced [63].

- Production of code or software that can be misused by third parties [63].
- Damage to the person due to entry errors [63–67].
- Public disclosure of confidential information about an individual [67].
- Publicity that places an individual in a bad light in public [67].
- Access to personal information when we don not realize it or with a purpose different from originary [63, 64, 66, 68].
- Use of disguised data collection devices that are difficult to recognize (ambient interfaces, wearables, ...).
- Use of copyrighted data or information [66]. Being unaware of the persuasion deliberately pursued through technology [69].
- Anticipated persistent and non-trivial effects in research participants [70].

Generally speaking, the measures to be taken to counter these possible risks involve ensuring the transparency in the criteria used by the system to create profiles; this warrants the users' awareness of ongoing data collection processes. This also makes the data collection process intelligible to the users and modifiable or interruptible at their will.

5   Defining Symbiotic Interaction

5.1   Core Proposition

Symbiotic interaction can be achieved by combining computation, sensing technology, and interaction design to realize deep perception, awareness, and understanding between humans and computers. Important aspects to implement are:

- Transparency, as the property of the computing to be for example accountable.
- Reciprocity and collaborative use of resources for both computers and humans.
- Symbiotic relationships are also characterized by goals and agency independence of humans and computers.

To address the above propositions, the research is highly multidisciplinary, involving computational science, engineering, psychology, and design. Specific branches of disciplines might provide decisive contributions, as recent work shows in neuroscience, neuropsychology, machine learning, and intelligent systems. While these basic scientific and engineering disciplines might provide theories and techniques to be adopted, a variety of human-oriented disciplines are needed for their applicability and real-world applications, such as for example design and anthropology. As we have seen, symbiotic interaction is related to and goes beyond recent frameworks, such as telepresence, affective and emotional computing, persuasive computing, and mixed initiative interaction. Over the years, the concept of symbiosis has surfaced from time to time in relation to human-computer Interaction. Different goals have inspired the rise. For example, this refers to magnifying human abilities through perceptual computing or the context of brain-computer interfaces. Current work in a variety of disciplines—including machine learning, sensing technologies, interaction design, neuroscience, and psychology—allows to create unprecedented relationships between computers and humans, characterized by deeper mutual understanding, cooperation, and independent agency. Such novel relationships can benefit from design and anthropological approaches to ensure transparency, optimal user experience, and appropriate considerations of ethical implications.

5.2   Taxonomy of Examples and Research Agenda

A first step can be to reflect on the different features that symbiotic interaction have compared to related frameworks. Figure 1 shows this comparison listing regarding understanding of context, modeling of cognitive state, emotion detection, experience induction, and how manipulable or accountable a system is. The comparison shows how symbiotic interaction could be seen as a comprehensive relationship with most features.

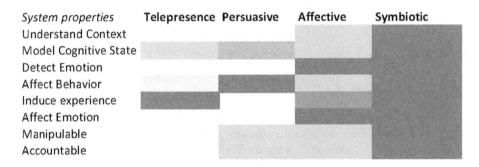

Fig. 1. Comparing system features for several frameworks related to Symbiotic Interaction

In addition, we can also present a taxonomy by taking concrete systems as examples and using the key dimensions of Symbiotic Interaction as axis namely (see Fig. 2)

- On the x-axis. The depth of understanding starts with no sensing, sensing only a location, recognizing activities (walking, running, biking, for example), cognition, language, and emotions. The hypothesis is that these are ordered by difficulty in this sequence.
- Transparency in the y-axis. Helps answer questions such as: is the system a black box, or provides transparent manifestation of what it is doing? Is it configurable? Is it reciprocal to the extent that the user can use resources of the system (computational constructs) and the system can use resources of the users (physiology, subliminal processes, history, etc.)?
- The z axis shows how independent the goals are. There can be only one individual goal that binds the user to the system, or there can be a variety of independent goals of different actors.

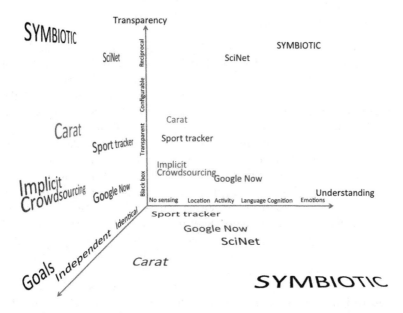

Fig. 2. Positioning of Symbiotic Interaction with examples of current systems. Examples used are Google Now, Carat [75] a crowdsourced batatrey management app, implicit crowdsourcing, SciNet [71,72], sport trackers

The diagram places examples systems on each of the 3 quadrants (x and y; y and z; and x and z). For example, in the x and y axes, Google Now has deeper understanding beyond physical positions and activities of the users and functions as a black box. SciNet, a system that provides access to computational construct of the search engine through interaction [71,72], is much higher on the transparency axis.

A sport tracker, which requires the user to select the activity that is being done, would be placed closer to the 0 axis. Moreover, a system like MatkaHupi [73,74] is a journey planner that automatically recognizes activities like walking, running, taking a bus, or driving a car; these activities would be placed further to the right in the x axis. As an example in the y and z axis quadrant a sport tracker has mostly one goal to support users in tracking their activity. Other systems might have multiple related goals. One example is Carat [75], in which users have an app for managing phone batteries. However, in doing so, users contribute data that is used in predicting how much battery time is saved by closing or updating applications. Here, the relation between the user and system is articulated with two interrelated goals; one is to help the user a save battery, and the other is to crowdsource data automatically for predicted use in the whole community of users. Finally, in implicit crowdsourcing, the goals can also be even more unrelated, and the user might not even know that a second goal is served. The diagram aims to show how symbiotic interaction would be placed in the corners of the quadrants in Fig. 2.

5.3   Some Further Challenges

While it is hard to draft a research agenda, we can attempt to summarize some elements for future research. We aimed at defining symbiotic interaction as a research program with distinctive characteristics, compared to previous human computer paradigms. While we have probably omitted some previous paradigms, we were able to single out some differences to previous paradigms as telepresence, affective computing, and persuasive technologies. Besides such paradigms, recent developments in adaptive interaction research are particularly interesting, considering some characteristics of symbiotic interaction, such as transparency and resourcefulness. For example, advanced visualizations that combine machine learning and user modeling have demonstrated how sense making can emerge by incrementally and interactively exploring a network of data [76]. Interactive visualization also supports user involvement in the recommendation by providing rationale behind suggested items [77] or visualizing relations of different queries and result sets [78]. Other works provide access to computational resources and constructs to manipulate and adapt the users' intent models [71]. A central open question focuses on if such approaches are to be seen as competing or alternative approaches to black box systems. These include conventional search interfaces, recent system that predict relevance of information through brain signals [79] or embodied conversational agent approaches that rely on natural dialogue understanding. These different approaches (resourceful systems vs. black box systems) might have fundamentally different assumptions on how the users should interact with the computing systems. On one hand, there is skillful user who is able to manipulate computing constructs, and on the other is a user who interacts with a humanoid system. Each approach is possibly most suited to certain types of tasks ad situations. Symbiotic interaction can follow these approaches as different research directions or investigate their combinations: (1) interactive adaptive visualizations as resourceful systems; or (2) natural interaction systems, including human robot interactions and embodied conversational agents.

A particularly interesting theme would be collective symbiotic systems: groups, cities, and societies. Here, goal independence of the symbiotic actants is taken at a collective level. It means that the actants (systems, users, and organizations) have a variety of different goals but can be made to interact for mutual benefits. As examples, consider crowdsourcing or implicit crowdsourcing.

6   Introduction to the Program

The program of Symbiotic 2014 contains a variety of papers that address the research from different perspectives. This paper aims to define symbiotic interaction and challenges for research, particularly comparing previous definitions of symbiotic interaction and related frameworks. Another paper by Lessiter et al. [80] discusses taxonomy of computer human interactions, particularly discussing implicit loops. Other papers present reviews of previous topics in implicit interaction regarding physiologically based relevance [81] or subliminal cueing [82]. Some work demonstrate applications of symbiotic interaction such as wearable robotic exoskeletons [83], or augmented reality in music browsing [84]. Some papers explore the experimentation with and inclusion of users as in the case of wearable technology acceptance [85] or use of pupil dilation information [86]. The program also offers a variety of posters and demos that present issues around interactive visualization [87], diversification in search [88], wearable haptics guidance [89], a multitouch interface to robots [90], balancing exploration and exploitation in image search [91].

Acknowledgments. This research has been partially funded by the by the European Community through the MindSee Project (FP7 – ICT; Grant Agreement 611570) and TEKES through the Re:Know project.

References

1. Licklider, J.C.R.: Man-computer symbiosis. IRE Trans. Hum. Factors Electron. **1**, 4–11 (1960)
2. Engelbart, D.C.: Augmenting human intellect: a conceptual framework (1962)
3. Roy, D.: 10^x—human-machine symbiosis. BT Technol. J. **22**(4), 121–124 (2004)
4. van Erp, J.B., Veltman, H.J., Grootjen, M.: Brain-based indices for user system symbiosis. In: Tan, D.S., Nijholt, A. (eds.) Brain-Computer Interfaces, pp. 201–219. Springer, London (2010)
5. Schalk, G.: Brain–computer symbiosis. J. Neural Eng. **5**(1), P1 (2008)
6. Barral, O., et al.: Covert persuasive technologies: bringing subliminal cues to human-computer interaction. In: Spagnolli, A., Chittaro, L., Gamberini, L. (eds.) PERSUASIVE 2014. LNCS, vol. 8462, pp. 1–12. Springer, Heidelberg (2014)
7. Aranyi, G., Kouider, S., Lindsay, A., Prins, H., Ahmed, I., Jacucci, G., Negri, P., Gamberini, L., Pizzi, D., Cavazza, M.: Subliminal cueing of selection behavior in a virtual environment. PRESENCE: Teleoperators Virtual Environ. **23**(1), 33–50 (2014)
8. Picard, R.W.: Affective computing: challenges. Int. J. Hum. Comput. Stud. **59**(1), 55–64 (2003)
9. Fogg, B.J.: Persuasive Technology: Using Computers to Change What We Think and Do. Morgan Kaufmann, Amsterdam (2003)
10. Reeves, B.: "Being there": television as symbolic versus natural experience. Stanford University, Institute of communication Research, Stanford Unpublished manuscript (1991)
11. Witmer, B.G., Singer, M.J.: Measuring presence in virtual environments: a presence questionnaire. Presence Teleoperators Virtual Environ. **7**, 225–240 (1998)
12. Lombard, M., Ditton, T.: At the heart of it all: the concept of presence. J. Computer-Mediated Commun. **3**(2) (1997)
13. Riecke, B.E., Von der Heyde, M.: Qualitative modelling of spatial orientation processes using logical propositions: interconnecting spatial presence, spatial updating, piloting, and spatial cognition. Technical Report No. 100, Max Planck Institute for Biological Cybernetics (2002)
14. Biocca, F., Harms, C., Burgoon, J.: Toward a more robust theory and measure of social presence: review and suggested criteria. Presence **12**(5), 456–480 (2003)
15. Spagnolli, A., Bracken, C.C., Orso, V.: The role played by the concept of presence in validating the efficacy of a cybertherapy treatment: a literature review. Virtual Reality **18**(1), 13–36 (2014)

16. Gamberini L., Spagnolli A.: Presence in mediated environments: accounting for new modes of human action. In: Lombard, M., Biocca, F., Freeman, J., IJsselsteijn, W., Jones, R. (eds.) Immersed in Media I: Telepresence Theory, Measurement and Technology. Springer, London (in press)

17. Mantovani, G., Riva, G.: Building a bridge between different scientific communities. On Sheridan's eclectic ontology of presence. Presence Teleoperators Virtual Environ. **10**, 537–543 (2001)

18. Spagnolli, A., Gamberini, L.: A place for presence. Understanding the human involvement in mediated interactive environments. PsychNology J. **3**(1), 6–15 (2005)

19. Gilroy, S.W., Cavazza, M.O., Vervondel, V.: Evaluating multimodal affective fusion using physiological signals. In: Proceedings of the 16th International Conference on Intelligent User Interfaces, February 2011, pp. 53–62. ACM (2011)

20. Ball, G., Breese, J.: Emotion and personality in a conversational agent. In: Cassell, J., Sullivan, J., Prevost, S., Churchill, E. (eds.) Embodied Conversational Agents, pp. 189–219. MIT Press, Cambridge (2000)

21. Rosis, F.D., Pelachaud, C., Poggi, I., Carofiglio, V., Carolis, B.D.: From Greta's mind to her face: modelling the dynamics of affective states in a conversational embodied agent. Int. J. Hum. Comput. Stud. **59**(1), 81–118 (2003)

22. Gamberini, L., Spagnolli, A., Corradi, N., Jacucci, G., Tusa, G., Mikkola, T., Zamboni, L., Hoggan, E.: Tailoring feedback to users' actions in a persuasive game for household electricity conservation. In: Bang, M., Ragnemalm, E.L. (eds.) PERSUASIVE 2012. LNCS, vol. 7284, pp. 100–111. Springer, Heidelberg (2012)

23. Spagnolli, A., Corradi, N., Gamberini, L., Hoggan, E., Jacucci, G., Katzeff, C., Broms, L., Jönsson, L.: Eco-feedback on the go: motivating energy awareness. Computer **44**(5), 38–45 (2011)

24. Japuntich, S., Zehner, M., Smith, S., Jorenby, D., Valdez, J., Fiore, M., Baker, T., Gustafson, D.: Smoking cessation via the internet: a randomized clinical trial of an internet intervention as adjuvant treatment in a smoking cessation intervention. Nicotine Tob. Res. **8 Suppl** 1(1), S59–67 (2006)

25. Consolvo, S., Klasnja, P., McDonald, D.W., Landay, J.A.: Goal-setting considerations for persuasive technologies that encourage physical activity. In: Proceedings of the 4th International Conference on Persuasive Technology, p. 8. ACM, April 2009

26. Chittaro, L.: Passengers' safety in aircraft evacuations: employing serious games to educate and persuade. In: Bang, M., Ragnemalm, E.L. (eds.) PERSUASIVE 2012. LNCS, vol. 7284, pp. 215–226. Springer, Heidelberg (2012)

27. Berdichevsky, D., Neuenschwander, E.: Toward an ethics of persuasive technology. Commun. ACM **42**(5), 51–58 (1999)

28. Smids, J.: The voluntariness of persuasive technology. In: Bang, M., Ragnemalm, E.L. (eds.) PERSUASIVE 2012. LNCS, vol. 7284, pp. 123–132. Springer, Heidelberg (2012)

29. Spahn, A.: And lead us (not) into persuasion..? Persuasive technology and the ethics of communication. Sci. Eng. Ethics **18**(4), 633–650 (2012)

30. Brynjarsdottir, H., Håkansson, M., Pierce, J., Baumer, E., DiSalvo, C., Sengers, P.: Sustainably unpersuaded: how persuasion narrows our vision of sustainability. In: Proceedings of the 2012 ACM Annual Conference on Human Factors in Computing Systems, pp. 947–956. ACM (2012)

31. Jameson, A.: Choices and decisions of computer users. In: Jacko, J.A. (ed.) The Human-Computer Interaction Handbook: Fundamentals, Evolving Technologies and Emerging Applications, 3rd edn. CRC Press, Boca Raton (2012)

32. Horvitz, E.: Principles of mixed-initiative user interfaces. In: Proceedings of SIGCHI, pp. 159–166 (1999)
33. Flemisch, F.O., Adams, C.A., Conway, S.R., Goodrich, K.H., Palmer, M.T., Schutte, P.C.: The H-Metaphor as a guideline for vehicle automation and interaction (2003)
34. Jameson, A.: Adaptive interfaces and agents. In: Sears, A., Jacko, J.A. (eds.) Human-Computer Interaction: Design Issues, Solutions, and Applications, p. 105. CRC Press, Boca Raton (2009)
35. Jordan, P.W.: Designing great stuff that people love. In: Blythe, M., Overbeeke, K., Monk, A., Wright, P. (eds.) Funology: From Usability to Enjoyment. Kluwer Academic Publishers, Dordrecht (2003)
36. McCarthy, J., Wright, P.: Technology as Experience. MIT Press, Cambridge (2004)
37. Suchman, L.: Plans and Situated Actions. CUP, Cambridge (1987)
38. Hollan, J., Hutchins, E., Kirsh, D.: Distributed cognition: toward a new foundation for human-computer interaction research. ACM Trans. Comput. Hum. Interact. (TOCHI) 7(2), 174–196 (2000)
39. Kuutti, K.: Activity theory as a potential framework for human-computer interaction research. In: Nardi, B. (ed.) Context and Consciousness: Activity Theory and Human-Computer Interaction, pp. 17–44. MIT Press, Cambridge (1996)
40. Carroll, J.M., Rosson, M.B.: Paradox of the Active User. The MIT Press, Cambridge (1987)
41. Bødker, S., Greenbaum, J., Kyng, M.: Setting the stage: workshops and group interaction. In: Greenbaum, J., Kyng, M. (eds.) Design at Work: Cooperative Design of Computer Systems, pp. 139–154. Lawrence Erlbaum Associates, Hillsdale (1991)
42. Ehn, P.: Work-Oriented Design of Computer Artifacts. Arbetslivscentrum, Falköping (1988)
43. Dourish, P.: Accounting for system behaviour: representation, reflection and resourceful action. In: Kyng, M., Mathiassen, L. (eds.) Computers and Design in Context, pp. 145–170. MIT Press, Cambridge (1997)
44. Dourish, P.: Where the Action Is: The Foundations of Embodied Interaction. MIT Press, Cambridge (2001)
45. Dourish, P.: Seeking a Foundation for context-aware computing. Hum. Comput. Interact. 16(2–4), 229–241 (2001)
46. Redström, J.: Designing everyday computational things. Ph.D. thesis, Department of Informatics, Gothenburg University, Gothenburg Studies in Informatics, No. 20 (2001)
47. Hallnäs, L., Redström, J.: Slow technology; designing for reflection. Pers. Ubiquit. Comput. 5(3), 201–212 (2001)
48. Svanæs, D.: Understanding interactivity, steps to a phenomenology of human-computer interaction. Doctoral Dissertation, Norwegian University of Science and Technology, Trondheim, Norway (1999)
49. Iacucci, G., Kuutti, K.: Everyday life as a stage in creating and performing scenarios for wireless devices. Pers. Ubiquit. Comput. J. 6(4), 299–306 (2002). (Springer-Verlag London Ltd.)
50. Macaulay, C., Jacucci, G., O'Neill, S., Kankaineen, T., Simpson, M.: The emerging roles of performance within HCI and interaction design. Interact. Comput. 18(6), 942–955 (2006). (Elsevier)
51. Binder, T., De Michelis, G., Ehn, P., Jacucci, G., Linde, P., Wagner, I.: Design Things. MIT Press, Cambridge (2011)

52. Jacucci, G., Spagnolli, A., Chalambalakis, A., Morrison, A., Liikkanen, L., Roveda, S., Bertoncini, M.: Bodily explorations in space: social experience of a multimodal art installation. In: Gross, T., Gulliksen, J., Kotzé, P., Oestreicher, L., Palanque, P., Prates, R.O., Winckler, M. (eds.) INTERACT 2009. LNCS, vol. 5727, pp. 62–75. Springer, Heidelberg (2009)

53. Laurel, B.: Computers as Theatres. Addison-Wesley, New York (1992)

54. Jacucci, C., Jacucci, G., Wagner, I., Psik, T.: A manifesto for the performative development of ubiquitous media. In: Proceedings of the 4th Decennial Conference on Critical Computing: Between Sense and Sensibility, pp. 19–28. ACM (2005)

55. Jacucci, G., Wagner, I.: Performative uses of space in mixed media environments. In: Turner, P., Davenport, E. (eds.) Spaces, Spatiality and Technology, vol. 5, pp. 191–216. Springer, Netherlands (2005)

56. Jacucci, G.: Interaction as performance: performative strategies in designing interactive experiences. In: Ekman, U., Bolter, J. D., Diaz, L., Engberg, M., Søndergaard, M. (eds.) Ubiquitous Computing, Complexity, and Culture. Routledge, New York (2015)

57. Ohm, P.: Broken promises of privacy: responding to the surprising failure of anonymization. UCLA Law Rev. **57**(6), 1701–1777 (2010)

58. El Emam, K., Neri, E., Jonker, E.: An evaluation of personal health information remnants in second-hand personal computer disk drives. J. Med. Internet Res. **9**(3), e24 (2007)

59. Tavani, H.T.: KDD, data mining, and the challenge for normative privacy. Ethics Inf. Technol. **1**, 265–273 (1999)

60. Tavani, H.T.: Genomic research and data-mining technology: implications for personal privacy and informed consent. Ethics Inf. Technol. **6**(1), 15–28 (2004)

61. van Wel, L., Royakkers, L.: Ethical issues in web data mining. Ethics Inf. Technol. **6**, 129–140 (2004)

62. Bennett, C.J.: Cookies, web bugs, webcams and cue cats: patterns of surveillance on the world wide web. Ethics Inf. Technol. **3**, 197–210 (2001)

63. Britz, J.: Technology as a threat to privacy: ethical challenges and guidelines for the information professionals. Microcomput. Inf. Manage. **13**(3–4), 175–193 (1996)

64. Papagounos, G., Spyropoulos, B.: The multifarious function of medical records: ethical issues. Methods Inf. Med. **38**(4–5), 317–320 (1999)

65. DeCew, J.W.: Privacy and policy for genetic research. Ethics Inf. Technol. **6**(1), 5–14 (2004)

66. Kuzu, A.: Problems related to computer ethics: origins of the problems and suggested solutions. Turkish Online J. Educ. Technol. **8**(2), 91–110 (2009)

67. Myers, M.D.: Ethical dilemmas in the use of information technology: an aristotelian perspective. Ethics Behav. **6**(2), 153 (1996)

68. Moor, J.H.: Towards a theory of privacy in the information age. ACM SIGCAS Comput. Soc. **40**(2), 31–34 (2010)

69. Naylor, J.: An analytical review of the experimental basis of subception. J. Psychol. **46**, 75–96 (1958)

70. Birgegard, A.: Persistent effects of subliminal stimulation: sex differences and the effectiveness of debriefing. Scand. J. Psychol. **49**(1), 19–29 (2008)

71. Ruotsalo, T., Peltonen, J., Eugster, Manuel, J.A., Głowacka, D., Konyushkova, K., Athukorala, K., Kosunen, I., Reijonen, A., Myllymäki, P., Jacucci, G., Kaski, S.: Directing exploratory search with interactive intent modeling. In: ACM International Conference on Information and Knowledge Management (2013)

72. Glowacka, D., Ruotsalo, T., Konuyshkova, K., Kaski, S., Jacucci, G.: Directing exploratory search: reinforcement learning from user interactions with keywords. In: Proceedings of the 2013 International Conference on Intelligent User Interfaces, pp. 117–128. ACM (2013)
73. Jylhä, A., Nurmi, P., Sirén, M., Hemminki, S., Jacucci, G.: Matkahupi: a persuasive mobile application for sustainable mobility. In: Proceedings of the 2013 ACM Conference on Pervasive and Ubiquitous Computing Adjunct Publication, pp. 227–230. ACM (2013)
74. Gabrielli, S., Forbes, P., Jylhä, A., Wells, S., Sirén, M., Hemminki, S., Nurmi, P., Maimone, R., Masthof, J., Jacucci, G.: Design challenges in motivating change for sustainable urban mobility. Comput. Hum. Behav. (2014)
75. Athukorala, K., Lagerspetz, E., von Kügelgen, M., Jylhä, A., Oliner, A.J., Tarkoma, S., Jacucci, G.: How carat affects user behavior: implications for mobile battery awareness applications. In: Proceedings of the 32nd Annual ACM Conference on Human Factors in Computing Systems, pp. 1029–1038. ACM (2014)
76. Chau, D.H., Kittur, A., Hong, J.I., Faloutsos, C.: Apolo: making sense of large network data by combining rich user interaction and machine learning. In: Proceedings of the SIGCHI Conference on Human Factors in Computing Systems, pp. 167–176. ACM (2011)
77. Verbert, K., Parra, D., Brusilovsky, P., Duval, E.: Visualizing recommendations to support exploration, transparency and controllability. In: Proceedings of the 2013 International Conference on Intelligent User Interfaces, IUI '13, pp. 351–362. ACM (2013)
78. Ahn, J.-W., Brusilovsky, P.: Adaptive visualization for exploratory information retrieval. Inf. Process. Manage. **49**(5), 1139–1164 (2013)
79. Eugster, M.J., Ruotsalo, T., Spapé, M.M., Kosunen, I., Barral, O., Ravaja, N., Jacucci, G., Kaski, S.: Predicting term-relevance from brain signals. In: Proceedings of the 37th International ACM SIGIR Conference on Research & Development in Information Retrieval (SIGIR '14), pp. 425–434. ACM, New York (2014)
80. Lessiter, J., Freeman, J., Miotto, A., Ferrari, E.: Ghosts in the machines: towards a taxonomy of human-computer interactions. In: Jacucci, G., Gamberini, L., Freeman, J., Spagnolli, A. (eds.) Symbiotic 2014. LNCS, vol. 8820, pp. 21-34. Springer, Heidelberg (2014)
81. Barral, O., Jacucci, G.: Applying physiological computing methods to study psychological, affective and motivational relevance. In: Jacucci, G., Gamberini, L., Freeman, J., Spagnolli, A. (eds.) Symbiotic 2014. LNCS, vol. 8820, pp. 35-46. Springer, Heidelberg (2014)
82. Negri, P., Gamberini, L., Cutini, S.: A review of researches on subliminal techniques for implicit interaction in symbiotic systems. In: Jacucci, G., Gamberini, L., Freeman, J., Spagnolli, A. (eds.) Symbiotic 2014. LNCS, vol. 8820, pp. 47-60. Springer, Heidelberg (2014)
83. Moreno, J.C., et al.: Symbiotic wearable robotic exoskeletons: the concept of the BioMot project. In: Jacucci, G., Gamberini, L., Freeman, J., Spagnolli, A. (eds.) Symbiotic 2014. LNCS, vol. 8820, pp. 72-86. Springer, Heidelberg (2014)
84. Åman, P., Liikkanen, L.A., Jacucci, G., Hinkka, A.: OUTMedia: symbiotic service for music discovery in urban augmented reality. In: Jacucci, G., Gamberini, L., Freeman, J., Spagnolli, A. (eds.) Symbiotic 2014. LNCS, vol. 8820, pp. 61-71. Springer, Heidelberg (2014)

85. Spagnolli, A., Guardigli, E., Orso, V., Varotto, A., Gamberini, L.: Measuring user acceptance of wearable symbiotic devices: validation study across application scenarios. In: Jacucci, G., Gamberini, L., Freeman, J., Spagnolli, A. (eds.) Symbiotic 2014. LNCS, vol. 8820, pp. 87-98. Springer, Heidelberg (2014)

86. Pluchino, P., Gamberini, L., Barral, O., Minelle, F.: How semantic processing of words evokes changes in pupil. In: Jacucci, G., Gamberini, L., Freeman, J., Spagnolli, A. (eds.) Symbiotic 2014. LNCS, vol. 8820, pp. 99-114. Springer, Heidelberg (2014)

87. Serim, B.: Querying and display of information: symbiosis in exploratory search interaction scenarios. In: Jacucci, G., Gamberini, L., Freeman, J., Spagnolli, A. (eds.) Symbiotic 2014. LNCS, vol. 8820, pp. 115-120. Springer, Heidelberg (2014)

88. Bandyopadhyay, P., Ruotsalo, T., Ukkonen, A., Jacucci, G.: Navigating complex information spaces: a portfolio theory approach. In: Jacucci, G., Gamberini, L., Freeman, J., Spagnolli, A. (eds.) Symbiotic 2014. LNCS, vol. 8820, pp. 138-144. Springer, Heidelberg (2014)

89. Hsieh, Y.-T., Jylhä, A., Jacucci, G.: Pointing and selecting with tactile glove in 3D. In: Jacucci, G., Gamberini, L., Freeman, J., Spagnolli, A. (eds.) Symbiotic 2014. LNCS, vol. 8820, pp. 133-137. Springer, Heidelberg (2014)

90. Andolina, S., Forlizzi, J.: A multi-touch interface for multi-robot path planning and control. In: Jacucci, G., Gamberini, L., Freeman, J., Spagnolli, A. (eds.) Symbiotic 2014. LNCS, vol. 8820, pp. 127-132. Springer, Heidelberg (2014)

91. Hore, S., Tyrväinen, L., Pyykkö, J., Glowacka, D.: A reinforcement learning approach to query-less image retrieval. In: Jacucci, G., Gamberini, L., Freeman, J., Spagnolli, A. (eds.) Symbiotic 2014. LNCS, vol. 8820, pp. 121-126. Springer, Heidelberg (2014)

Ghosts in the Machines: Towards a Taxonomy of Human Computer Interaction

Jane Lessiter[(✉)], Jonathan Freeman,
Andrea Miotto, and Eva Ferrari

Psychology Department, Goldsmiths,
University of London, London, UK
{J.Lessiter,J.Freeman,A.Miotto,E.Ferrari}@gold.ac.uk

Abstract. This paper explores a high level conceptualisation (taxonomy) of human computer interaction that intends to highlight a range of interaction uses for advanced (symbiotic) systems. The work formed part of an EC-funded project called CEEDs which aims to develop a virtual reality based system to improve human ability to process information, and experience and understand large, complex data sets by capitalising on conscious and unconscious human responses to those data. This study, based on critical and creative thinking as well as stakeholder consultation, identified a range of variables that impact on the types of possible human computer interaction, including so called 'symbiotic' interactions (e.g., content displayed – raw/tagged; user response – explicit/implicit; and whether or not there is real time influence of user response on content display). Impact of variation in the number of concurrent users, and of more than one group of users was also considered. This taxonomy has implications for providing new visual stimuli for creative exploration of data, and questions are raised as to what might offer the most intuitive use of unconscious/implicit user responses in symbiotic systems.

Keywords: Symbiosis · Symbiotic · Human-computer · Interaction · Technology-mediated · Taxonomy · Classification · Presence · Interaction design · Affective systems · Intuitive interaction · Attention · Mediated · Displayed · Environment · Confluence

1 Introduction

The quality and forms of technologies to support interactions between people and computers have markedly increased in recent years, from simple pointers, navigators and button click inputs to more natural and intuitive, even immersive, controls. Interactions with technologies can be experienced as an extension of the self, typically through increased experience/practice, for example, driving a car. However newer technologies are attempting to remove the 'translator' in the mechanics and plug more directly into our conscious as well as unconscious perceptions and intentions offering the potential for a truly seamless, transparent experience.

Intuitive interactions enable increased focus and immersion in the task to hand, reducing the division in attention between the primary (target) and secondary (machine interaction) tasks. Indeed the term 'presence' has been used to describe this subjective

© Springer International Publishing Switzerland 2014
G. Jacucci et al. (Eds.): Symbiotic 2014, LNCS 8820, pp. 21–31, 2014.
DOI: 10.1007/978-3-319-13500-7_2

sense of 'being there' in a mediated/displayed environment (e.g., Barfield et al. 1995) and is often used to measure perceived quality of experience of a displayed environment (e.g., Lessiter et al. 2001). In these contexts, distractions emerging from the 'real world' can include clumsy and awkward system interactions that demand user attention and break the spell to mediated presence. Identifying ways of minimising distractions or at least their impact on both the user and the machine is important to the fluidity, meaningfulness and accuracy of inferred experience.

A high level conceptualisation of human computer interaction would benefit our understanding of the usage possibilities within this new genre of what might be called 'symbiotic systems'. This paper describes one such interaction taxonomy informed by consultation with stakeholders, and critical and creative thinking. The work was born out of an European Commission (FP7 FET) funded project called CEEDs – the Collective Experience of Empathic Data systems - which explores the potential for both conscious and unconscious (multi-)user control over displays of complex data.

2   Classifying Human-Computer Interactions

The term 'taxonomy' originated in the biological sciences. Taxonomies have now extended to any "ordered classification, according to which theoretical discussions can be focused, developments evaluated, research conducted, and data meaningfully compared" (Milgram and Kishino 1994, p. 1323). Such classifications can shed light on the critical and optional parameters that can serve to enhance or reduce the quality of experience of that human computer interaction.

There are taxonomies for various types and aspects of human-computer systems which can make it difficult to compare and pool these specific taxonomies as the extent to which they overlap is unclear (von Landesberger et al. 2014).

From a more general perspective, Parasuraman et al. (2000) provided a framework for dynamic function allocation systems (following the traditional trajectory of HCI) comprising a dimension of automation ranging from fully automated to fully manual for each of four categories of function: (a) information acquisition; (b) information analysis; (c) decision and action selection; and (d) action implementation. However, this type of classification does not clearly address systems which may not seek to replace effort on behalf of the user, but rather complement or harness human efforts at different levels of our awareness as may be the case with systems that aim to be symbiotic.

3   Human-Computer Symbiosis

Symbiosis traditionally denotes the relationship between two or more organisms 'living together' in a mutually advantageous manner. In an HCI context it similarly suggests a mutually complementary understanding between entities (e.g., Licklider 1960; Grootjen et al. 2010; García Peñalvo et al. 2013) that builds and grows with each interaction. Indeed, relationships that are truly symbiotic are expected to be experienced as natural, fluid and intuitive.

How this might be achieved raises questions about how we make sense of ourselves and how we believe that systems will be able to understand and make sense of us. Consequently, what user data and processes (e.g., sequences of action) are important and what could be inferred from them? Further, in what contexts or scenarios could these symbiotic relationships with computers be of particular advantage? And importantly, how might users experience symbiotic relationships with computers, given that this is a new type of human-computer interaction? Theoretically we imagine it will feel seamless and natural but until these systems are tried, tested, technologically optimised, and users become more familiar with their implementation, it is unclear how users will initially experience and respond to them.

There are numerous projects that have developed or are developing applications of symbiotic systems that aim to improve *relevance* of system outputs to user requirements, for instance of literature searches (e.g., EC MindSee: see www.mindsee.eu), product recommendations (e.g., Adidas-Intel's 'adiVerse'), games (e.g., Kinect 2 and Valve, both of which make use of physiological responses to influence game play), and vehicles (e.g., AutoEmotive).

Applications are naturally variable in the quality of the interaction/symbiosis depending on the validity (meaningfulness) and reliability (accuracy vs. uncertainty and consistency) of the user data that are selected, how inferences are made by the system and the timeliness of the system's response. In today's world, massively multivariate data are being collected at an unfathomable rate and yet are neither fully understood nor exploited for advantageous benefits to ourselves and wider society. Of course human-computer interaction quality is currently dependent on, and is limited by, what we can tell machines about ourselves and how to learn about us. We are limited by our sensory, perceptual and cognitive abilities, and by the engineered tools we have creatively constructed throughout our evolution to support our capacity to understand the world. But what if new technology can help us exploit previously unexplored territories that reside in our bodies as extensions of our interaction with the world? Eventually machine learning and its application in symbiotic systems may supersede our current creative capacity in as early as 2021 as predicted by Vinge (2006) (cited in Grootjen et al. 2010).

3.1 Symbiosis in Context: The CEEDs Project

The Collective Experience of Empathic Data Systems (CEEDs) project commenced in 2010 and aims to address the data deluge problem which includes symbiotic man-machine interactions. It proposes to develop a virtual reality based system to improve human ability to process information, and experience and understand large, complex data sets by capitalising on conscious and unconscious human responses to those data (see: ceeds-project.eu; e.g., Lessiter et al. 2011; Pizzi et al. 2012; Freeman et al. (2015)).

Central to the symbiosis is having the interaction between representations of two entities that have intent to make sense of the other, mediated by a 'sentient' autonomous agent. The representation of the 'human' (to the computer) relates to the physical, observable manifestations of the user's self that the computer is watching (user responses). The computer represents its 'self' through sensorial information to be solved (the task related data – the data deluge) that is amenable to human detection (visualisation e.g. patterns) at varying levels of awareness. This is how the entities exchange information and the system and user become "confluent into a symbiotic cooperation" (van Erp et al. 2010, p. 202).

The CEEDs system is a type of dynamic/adaptive visual analytic as it primarily represents its understanding of the user's intentions to explore the data visually: what patterns do users notice/respond to and why? And how should the system appropriately respond to support increased user understanding, discovery and creativity? Such systems capitalise on human cognition (e.g., perception and decision making) and machine based data processing, analysis and learning, providing a bridge between exploration and analysis (Endert et al. 2014). How quickly the system responds to user interactions can also pose a technical hurdle to the fluidity and perceived symbiosis of the experience.

In the early stages of the CEEDs project, development of use cases and scenarios was required across a wide range of potential application domains: big visualised data in the areas of neuroscience, history, archaeology and design/retail (commerce). To this end, system commonalities across application scenarios were sought based around possible ways in which users' responses may be used as inputs and outputs in any CEEDs application. A taxonomy or classification system of human computer interaction that included symbiotic data exchange was required.

4   Method

To inform the development of the taxonomy of interaction uses, stakeholders working within each of the target application domains were consulted. Along with pragmatic questions relating to the types of data used in their big data field, they were asked to (qualitatively) comment on the relevance of the initial aims of CEEDs, as stated in the project's Description of Work document for their specific application area, and to suggest other goals they envisaged being met by CEEDs.

The objective was to gather as many user requirements as possible which would provide a wide range of material to refine and elaborate a smaller selection of higher level uses that were deemed valid, symbiotic, in-scope of the project and could be developed as prototypes. Having a wide application remit was useful for developing a taxonomy of interaction uses that was general enough to apply across different contexts.

Replies were received from seven stakeholders around half of whom were project partners. Responses included five university departments, one historical museum/ memorial site and a 'white goods' manufacturer. The majority were academics.

Critical and creative thinking was used to identify underlying processes and/features that may vary in one type of interaction experience to another.

5   Results

The contextualized potential applications of CEEDs derived from the Description of Work and stakeholder feedback were explored for interaction styles and themes. Across the application areas, there was also some broad consistency in the goals that CEEDs technology could support. For instance, CEEDs supports insight and adaptability to users' responses to data which makes it a useful tool for the following interrelated uses:

- discovering unknown relationships (e.g., between user responses and stakeholder data i.e., adding metadata to stakeholder databases)
- personalising experiences (e.g., refining choices)
- validating relationships (e.g., best fit)
- representing relationships (e.g., reviewing data)
- optimising experiences to a given construct (e.g., influencing others, learning sequences of actions, improving memorability of information, optimising enjoyment/presence).

Some stakeholder goals/requirements indicated a distinction between (a) (primary) CEEDs end users – users/interactors and (b) (secondary) CEEDs beneficiaries – CEEDs system data owners. (Primary) CEEDs end users are those who use and interact with the system. For instance, customers are supported in their product choices by CEEDs offering a personalised service based on their own (stored and/or real time) unconscious desires and preferences. As an alternative example, consider a team of neuroscientists attempting to validate/refute models to explain patterns of data. They are supported in this discovery process by CEEDs technology because it harnesses their unconscious responses to different visualisations of those models with the data. The neuroscientists can test these models for unconscious 'goodness of fit'. Primary CEEDs end users could be both expert/professional users as well as novices.

Other stakeholder goals suggested that some CEEDs users could be more correctly classified as CEEDs beneficiaries as they are (secondary) CEEDs users of others' data. These are characterised as CEEDs system/database owners and can analyse end user responses to data in all sorts of ways. Beneficiaries could use CEEDs user data to optimise displays for different goals (e.g., learning, empathy, sales); predicting and influencing a users' behaviour by understanding their states/plans/intentions in a given context. For instance, design teams may be beneficiaries if they explore their customers' implicit reactions to products to improve product design. Most users in this category were experts/professionals.

5.1 Components of Symbiotic Human-Computer Interaction

A taxonomy of these different types of human computer interaction/symbiosis within a CEEDs context was developed that identify a number of dimensions or factors which may change from one experience to another.

Three main entities were considered relevant in the context of inputs and outputs in the human computer interaction, namely: the user, the 'CEEDs engine' (a sentient autonomous agent/computer), and the content/data display through which the computer relates to the user (see Fig. 1).

The taxonomy assumes that any data displayed will have meaning. There are two main sources of data: the raw data (that comprise the data deluge) and response data from the user. The raw data (before any users have experienced and responded to it) is visualised, potentially represented in other modalities, and 'produced' or contextualised before it is presented to a user: the raw data will not appear as the database itself.

In this conceptualisation of symbiotic interaction, user response data (to the presentation of the raw data) are recorded and annotated into a copy of the raw dataset. For instance data recorded may include what the user response was (e.g., a smile, a vocal command, pupil dilation), to what specifically in the display, for how long the user response lasted, and the inferred meaning (interpretation) of the single/combined user response(s). This process of recording and storing user response data is termed 'tagging' here.

The taxonomy includes variation in (a) the nature of the content displayed (raw or pre-tagged data), (b) whether or not current users' implicit/explicit data are measured/monitored, and (c) whether or not there is any real time feedback from the system to user responses (in whatever form that feedback might take). Additionally there are example representations of multiple users in any session. Varying these characteristics produces a wide range of interaction possibilities, the outputs of some of which are difficult to imagine and remain to be explored from a user experience perspective. Crucially, the displays of those datasets (raw and user) can be combined in different ways to produce novel outputs that users can either view and/or actively explore. For further information about the core features of the CEEDs system, see Freeman et al. (2015).

In the pictorials that were developed to accompany the taxonomy (nb. The full set of pictorials are beyond the scope of this position paper), the three entities are contained within a larger boxed space indicating the 'current session'. In its simplest use, the content display reflects variation in the task related dataset to be explored (raw data). But with no feedback from the user, this provides a passive mode of operation (merely 'viewing') akin to television watching (see Fig. 1).

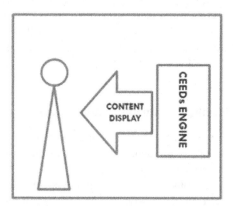

Fig. 1. Three components of human computer interaction: passive viewing

5.2 Active and Passive Interactions

In a fully interactive technology-mediated experience, the content display presents variation in both the raw data, and the CEEDs engine's real time reflection of its understanding of the current user (the user dataset/model). Its understanding develops with each interaction by the user with the data displayed.

Of course, there are also many possible viewing experiences that are passive for the current user whereby they have no real time explicit or implicit control over the content display. For instance, in a passive mode, it is possible that (a) the current user response data is nevertheless being collected albeit not represented whilst 'viewing' and/or (b) the data displayed to the current user might already be an amalgamation of raw data and user data (e.g., experiencing someone else's visual experience of the data, or indeed one's own experience from a previous session). These technology mediated experiences, whilst passive, may also provide visual stimuli that inspire insightful explorations of data. The style of interaction between the human user and computer (CEEDs engine) via the content display is therefore related to the presence or absence of real time or delayed explicit and implicit user response feedback to the system.

5.3   Types of User Responses

The more the user is physically immersed and feels present in the displayed experience, the more accurate the user model that can develop. User responses collected by the system are ideally in response to what is displayed and not caused by extraneous uncontrolled events outside of the displayed world. This underlines the importance of the fluidity and usability of the user's explicit means of controlling the display particularly with regard to naturalness of (expected) interaction.

'Explicit' (user) responses refer here to overt, deliberate 'conscious' responses. These include behavioural responses such as gesture, pointing, verbal responses/ speech, button pressing, manipulation of tangible representations and motion/trajectory. It is expected that the user's sense of symbiosis will increase with improvements in the accuracy and attentiveness of the CEEDs engine and naturalness of expected response of the display (output) to users' explicit responses (inputs). This relies on the system understanding what users consider appropriate and expected. This area is generally well understood in HCI.

In contrast, user responses are called 'implicit' when they refer to covert, uncontrolled responses that are 'unconscious'. These could be physiological (e.g., ECG, respiration, EDR, EEG, EMG, pupil dilation) or behavioural (e.g., blink rate, eye-tracking, reflexive postural and physical responses, vocal emotion). Unlike explicit user responses, it is not yet clear how users might expect the display to 'intuitively' respond to their own implicit responses to the content display and indeed how this impacts on subsequent explicit and implicit action loops. How will we make sense of being faced with our own responses that previously lay beneath our conscious awareness?

5.4   Key Variables in Human-Computer Interactions

In this initial taxonomy of human-computer (symbiotic) interactions, user responses that are implicit or explicit can be relevant, or not, to any particular CEEDs session. User response feedback in the current session can be real time or derived from a previous interaction with the same or a different user's response data. Thus user response data can be collected without having real time impact.

Where a previous user session's data is relevant, data displayed in the current session have 'pre-tags' based on those responses of users from previous sessions (which may or may not include previous experiences of the current user with that raw data). It may be possible for a current user to be presented with pre-tagged raw data (derived from a previous session) as well as real time tags based on the current users' responses. For instance, a designer may wish to explore using explicit responses (e.g., gesture) the data of a target user who has previously explored the latest of their product designs to understand which aspects of the product were particularly noticeable and liked by the user.

5.5   Passive Interactions and Example Uses

Passive interaction examples are those in which the user cannot influence the display in real time (no user feedback of explicit and/or implicit responses). However, it is possible that the current user's implicit and/or explicit response data are nevertheless being captured and tagged on to the raw dataset. This tagged data could then be explored by the same/another user at a different session. An example of this type of scenario would be circumstances where it is important to control presentation of (inert/moving) stimuli so that multiple users' responses to standardised (identical) stimuli can be collected and later reviewed.

In this type of 'review' scenario, a previous user's data are 'overlaid', tagged or have influenced the way in which raw data is presented. The current user's responses to those tagged raw data are inconsequential if their responses are not being collected.

In another scenario the current user could experience 'tagged' raw data, and whilst their own explicit and/or implicit responses may be collected and stored, they may have no real time influence on the display. This could represent a current user's experience of data that has been optimised for a given construct (e.g., learning) based on a previous user's data and for which the beneficiary (e.g., an expert) is testing the effectiveness of this representation with the current user.

5.6   Active Interactions and Example Uses

There is a range of potential active interaction experiences. In all examples, raw data is influenced in real time by the users' own implicit and(/or) explicit responses. This relies on the computer (in this instance, the CEEDs engine) to develop a user model based on user responses to data displayed. Its user model influences the display to support the goals of that experience (e.g., learning, maintain a particular level of arousal).

A user could have real time influence over their experience of pre-tagged stakeholder data. This scenario is relevant where the goal is to reinforce or strengthen associations between multiple serial/concurrent users' responses and the representations in the stakeholder dataset.

Any CEEDs active interaction session could take place with single or with multiple concurrent users. For instance, multiple concurrent users could provide their own responses to pre-tagged raw data. These possibilities raise issues about how the computer will deal with user data from multiple simultaneous inputs. For instance, a teacher in training their students to look for significant patterns in raw data may require that the computer weights the group responses (and thus, the influence on the display) to that expert's response data.

More complex human-computer symbiotic interactions are possible. For instance two groups of remotely located users (explorers and evaluators) could interact with the same dataset: evaluators observe in real time how the explorers react to the data. If evaluators' explicit responses are weighted in their favour they are given more control over the display through their explicit reactions. In the CEEDs system, members from both groups are considered as part of one group (only the applied weighting/roles will vary). An example might be that the evaluators are a product design team that has some explicit control over how the explorers are experiencing a product. For instance, they might (explicitly) command the system to direct the explorers' attention in real time to a new design feature to better understand how users respond to it.

There are further possible types of very complex interactions to be explored using this taxonomy. A flavour of what might be possible has been provided in the examples considered here. Whilst this taxonomy served the purpose of supporting the definition of what is a CEEDs experience, its generality to a wide range of other interactive systems remains to be tested. Further, as highlighted by Nickerson et al. (2013), taxonomies of information systems that have been conducted often fail to apply a rigorous methodology in their development. This paper faces this criticism. Nickerson et al. have recently provided methodological guidance based on other taxonomic literature in other research fields. Future exploration of interaction taxonomies might compare the classification system outlined in this paper with those derived from more rigorous approaches.

6 Conclusions

This paper provides an initial thought based investigation towards a taxonomy of human-computer interaction that includes advanced (symbiotic) systems. The work was inspired through an EC funded project called CEEDs to research and develop a visual analytics based system that facilitates user creativity and discovery in massive datasets by exploiting measurable aspects of human perception and intention.

Through critical and creative thinking and stakeholder consultation, a range of variables were identified that influence what might be possible with new systems that seek to offer symbiotic and other types of human-computer interaction: primarily, content displayed (raw/tagged), types of user responses stored by the system (explicit/ implicit), and real time influence (of explicit/implicit responses). Impact of variation in the number of concurrent users, and of more than one group of users was also considered.

It is possible that some types of so called symbiotic systems may not be perceived by users as symbiotic. People are variably attuned to the (conscious/unconscious) impacts they have on others (people/objects) in our day to day interactions with the world. However people as technology users are far less familiar with being exposed through technology to the influence of their implicit responses to the world (physical or virtual) around them. Understanding how to optimally harness this symbiosis for different types of interaction to facilitate creativity, discovery and understanding of complex data is no minor task. How do different levels of self-awareness of one's own implicit responses impact on the sense of this kind of symbiotic 'attunement'? What factors influence the extent to which the interactions derived from implicit user responses are perceived as natural and not jarred?

Attempts at symbiotic interactions that make use of implicit user responses may initially require practice and experience, like driving a car. Perhaps as we become more familiar with using these new symbiotic tools, we will learn new ways of making more integrated sense of our technology reflected intentions.

Acknowledgements. This work was conducted within the European Commission's Future and Emerging Technologies CEEDs project, a 48 month project part funded under FP7 and commenced in September 2010 (Project number: 258749; Call (part) identifier: FP7-ICT-2009-5).

References

Barfield, W., Sheridan, T., Zeltzer, D., Slater, M.: Presence and performance within virtual environments. In: Barfield, W., Furness, T. (eds.) Virtual Environments and Advanced Interface Design. Oxford University Press, Oxford (1995)

Endert. A.. Shahriah Hossain, M., Ramakrisnan, N., North, C., Fiaux, P. Andrews, C.: The human is the loop: new directions for visual analytics. J Intell. Inf. Syst. (2014). http://people. cs.vt.edu/naren/papers/jiis-human-is-loop.pdf. Accessed 17 July 2014

Freeman, J., Miotto, A., Lessiter, J., Verschure, P., Omedas, P., Seth, A.K., Papadopoulos, G.T., Caria, A., André, E., Cavazza, M., Gamberini, L., Spagnolli, A., Jost, J., Kouider, S., Takács, B., Sanfeliu, A., De Rossi, D., Cenedese, C., Bintliff, J.L., Jacucci, G.: The Human as the mind in the machine: addressing big data. In: Gaggioli, A, Ferscha, A,. Riva, G, Dunne, S, Viaud-Delmon, I. (eds.) Human Computer Confluence: Advancing our Understanding of the Emerging Symbiotic Relation between Humans and Computing Devices. Versita (2015)

García Peñalvo, F.J., Colomo Palacios, R., Hsu, J.Y.J.: Discovering knowledge through highly interactive information based systems Foreword. Journal of Information Science and Engineering, 29(1), (2013). http://gredos.usal.es/jspui/bitstream/10366/121169/1/DIA_ GarciaPenalvo_DiscoveringKnowledgeThroughHighly.pdf

Grootjen, M., Neerincx, M.A., van Erp, J.B.F., Veltman, J.A.: Human-computer symbiosis by mutual understanding. In: CHI 2010, Atlanta, Georgia, USA. http://www.eecs.tufts.edu/~agirou01/ workshop/papers/grootjen-CHI2010-BrainBodyBytes2010.pdf. Accessed 17 July 2014

Lessiter, J., Freeman, J., Keogh, E., Davidoff, J.: A cross-media presence questionnaire: the ITC-sense of presence inventory. Presence Teleoperators Virtual Environ. **10**(3), 282–297 (2001)

Lessiter, J., Miotto, A., Freeman, J., Verschure, P., Bernardet, U.: CEEDs: Unleashing the power of the subconscious. Procedia Comput. Sci. **7**, 214–215 (2011) (Proceedings of the 2nd European Future Technologies Conference and Exhibition 2011: FET 11)

Licklider, J.C.R.: Man computer symbiosis. IRE Trans. Hum. Fact. Electron. **HFE-1**, 4–11 (1960)

Milgram, P., Kishino, F.: A Taxonomy of mixed reality visual displays. IEICE Trans. Inf. Syst. **HFE-1**, 1321–1329 (1994)

Nickerson, R.C., Varshney, U., Muntermann, J.: A method for taxonomy development and its application in information systems. Eur. J. Inf. Syst. **22**, 336–359 (2013)

Parasuraman, R., Sheridan, T.B., Wickens, C.D.: A model for types and levels of human interaction with automation. IEEE Trans. Syst. Man Cybern. Part A Syst. Hum. **30**(3), 286–297 (2000)

Pizzi, D. Kosunen, I., Viganó, C., Polli, A.M., Ahmed, I., Zanella, D., Cavazza, M., Kouider, S., Freeman, J., Gamberini, L., Jacucci, G.: Incorporating subliminal perception in synthetic environments. In: Proceedings of the ACM Conference on Ubiquitous Computing, pp. 1139–1144 (2012)

van Erp, J.B.F., Veltman, H.J.A., Grootjen, M.: Chapter 12: Brain based indices for user system symbiosis. In: Tan, D.S., Furness, T. (eds.) Brain-Computer Interfaces: Applying Our Minds to Human-Computer Interaction. Springer, London (2010)

von Landesberger, T., Fiebig, S., Bremm, S., Kuijper, A., Fellner, D.W.: Interaction taxonomy for tracking of user actions in visual analytics applications. In: Huang, W. (ed.) Handbook of Human Centric Visualization, pp. 653–670. Springer, New York (2014). http://link.springer.com/chapter/10.1007/978-1-4614-7485-2_26

Reviews of Implicit Interaction

Applying Physiological Computing Methods to Study Psychological, Affective and Motivational Relevance

Oswald Barral$^{(\boxtimes)}$ and Giulio Jacucci

Department of Computer Science, Helsinki Institute for Information
Technology HIIT, University of Helsinki, Helsinki, Finland
{oswald.barral,giulio.jacucci}@helsinki.fi

Abstract. Relevance in information science has been studied for over forty years and robust frameworks have been derived. However, information retrieval systems are still using mainly objective, algorithmic measures of relevance. The aim of the present paper is to raise a discussion around the possibility that bring state-of-the-art physiological computing methods to model subjective components of relevance. We center the discussion on the relevance types known in the information science literature as psychological, affective and motivational relevance. The paper presents a definition of these concepts, as well as an overview of the recent advances in physiological computing methods developed in information science and information retrieval. We conclude with a discussion around the potential of physiological computing methods to model psychological, affective or motivational relevance.

Keywords: Physiological computing · Affective relevance · Psychological relevance · Information retrieval

1 Introduction

The concept of relevance has been widely studied in the field of information science and a strong theoretical background has been developed around it [1]. It is recurrent to find distinctions between objective relevance, which is intrinsically dependent on the item to evaluate; and subjective relevance, which is dependent on the perception that the user has of such item. Information systems, namely information retrieval systems, are commonly based on objective, algorithmic relevance metrics. These metrics are especially well suited as they are easily quantifiable measures of relevance. Nevertheless, objective relevance alone is far from indicating the real relevance of an item in a given context. It is evident that subjective relevance assessments play a strong role in the success of an information-seeking task. Subjective relevance has multiple facets and, in the present paper, we aim to engage a discussion around the concepts of psychological, affective and motivational relevance. These concepts address the relationship between the user's current emotional and cognitive state and the

© Springer International Publishing Switzerland 2014
G. Jacucci et al. (Eds.): Symbiotic 2014, LNCS 8820, pp. 35–46, 2014.
DOI: 10.1007/978-3-319-13500-7_3

information item. By studying these components, we tackle directly the information need and its relationship between the user's current emotional, cognitive and affective state. To give a simple example, during a complex information-seeking task, an item can be perceived as relevant when the user is engaged into the search task, but perceived as irrelevant when the user feels overwhelmed, frustrated or bored.

It is not the aim of this paper to review relevance in information science, but to compile pertinent concepts regarding subjective relevance as well as review state-of-the-art research on physiological computing in information retrieval. By doing so, we expect to bring arguments to raise a discussion around the need to apply physiological computing techniques to model affective, motivational and psychological relevance, in order to enhance the performance of current information retrieval systems in complex search scenarios. We believe that such discussion will benefit future information retrieval systems, towards more symbiotic human-system interactions [2].

2   Subjective Relevance in Information Science

Relevance has been theoretically analyzed over the past forty years in the field of information science by researchers such as Saracevic [3–7], Schamber [1], Mizzaro [8,9], Ingwersen [10,11], and Borlund [12] among others.

Saracevic [3] carried out the first comprehensive review in information science around the concept of relevance back in 1975. He compiled relevant work mainly from the two previous decades, addressing relevance from philosophy all the way to relevance in information retrieval systems. Additionally he aimed at providing a framework to study relevance within the field of information science. Later on, Saracevic updated his review as the field was growing and the discussion was enlarging [6,7]. However, Saracevic has not been the only researcher to study the multifaceted aspects of relevance. Schamber [1] or Mizzaro [8], are examples of authors that made a substantial contribution during the 90's by writing comprehensive and historical reviews on relevance. We encourage interested readers to go through their work in order to gain a deeper insight in the history of relevance in information science.

A large consensus exists within the community in considering relevance as multifaceted, dynamic and with a high subjective component [5,9–13]. In the following sections, we will tackle some of the subjective aspects of relevance, leaving out of the analysis objective and algorithmic measures of relevance. Specifically, we will go through three relevance components addressed in the literature by different authors, that are highly related and interconnected, and which we believe can be studied together. These components define the relationship between the user state, the information need and the information item.

2.1   Psychological Relevance

The concept of psychological relevance was first raised in cognitive sciences, in the popular book "Relevance: Communication and Cognition" by Sperber and Wilson [14]. In the book, the authors define a framework for relevance in communication and cognition theory, based on the assumption that human cognition only pays attention to potentially relevant information. They argue that relevance is framed in the cognitive system according to comparisons between assumptions and a particular cognitive state. In 1992, based on their theoretical framework, and borrowing some of their theoretical definitions, Harter defined psychological relevance applied to the field of information science [15]. He used the concept of *relevance of a phenomena*, elaborated in the above-mentioned book by Sperber and Wilson, as a mainstay for his definition. In plain words, this concept refers to the fact that phenomena have a direct impact on the person's cognitive state, and therefore affect his assumptions by making them stronger or more manifest.

Harter defined psychological relevance in information science as follows: *"Psychological relevance allows us to talk about an 'information need' as the current context –the cognitive state at a given time– of an individual who consults an information system"* [15]. In his work, he illustrates through examples how the cognitive state of an individual has a direct impact on the perceived relevance of information items. Moreover, he makes a distinction between relevance and *"weak relevance"*. In Harter's view, relevance occurs when the fact of accessing an information item implies a direct modification of the user's cognitive state. However, *weak relevance* occurs when the user anticipates a change on his cognitive state, even though accessing the information item does not imply a direct modification of the cognitive state. Other variants of the concept of psychological relevance are well illustrated by Harter through well-elaborated examples.

Harter's definition has been present in the work of other researchers, mainly when reviewing the literature, but it has not had so far a direct implication in the implementation of information retrieval systems. In the following paragraph, we would like to cite other authors when rephrasing Harter's definition of psychological relevance. By doing so, we hope the reader will gain a clearer understanding of the concept.

In her review, Schamber [1] rephrased Harter's concept of psychological relevance as follows: *"Psychological relevance assumes that users are actively seeking information that will change their internal context or store of knowledge, and that if an information item does this, users will perceive it to be relevant"*. Saracevic [4] pointed at Harter's definition in the following terms: *"Psychological relevance is viewed as a dynamic, ever changing interpretation of information need in relation to presented texts. It is based on an assumption (stated as fact) that the 'searcher's cognitive state changes and evolves with the discovery of each relevant citation'."* In fact, Saracevic proposed the term *"cognitive relevance"* for the concept, as he argued that the focus was on cognition, rather than on psychology. Finally, to summarize, it is pertinent to cite Mizzaro's reference to Harter's work [8]: *"Harter (1992) applies the theory of psychological relevance, proposed by Sperber and Wilson, to the concept of relevance in information science. He obtains an elegant framework and draws some very interesting conclusions for IR and bibliometrics."*

2.2   Affective and Motivational Relevance

As we just discussed, psychological relevance describes the relationship between the users' prior knowledge and cognitive state, and the information item. Instead, affective or motivational relevance is defined as the relationship between the users' intents goals and motivations, and the information item [4,11]. There-fore, affective or motivational relevance is goal oriented, and is dependent on the user's current cognitive context [13]. Saracevic [4,5] first adapted this term from Schutz's concept of relevance in philosophy [16]. Schutz named *"motivational relevance"* as one of the three basic types of relevance, together with *"topical relevance"* and *"interpretational relevance"*, that he defined when studying human relationships in the social world. According to Schutz, this kind of relevance is the one that defines the course of action, the action to be executed, the selection of a specific alternative. Saracevic extends and applies the notion to the field of information science. Moreover, he couples it with the notion of *affective relevance*, resulting in the following definition: *"Motivational or affective relevance is the relation between the intents, goals, and motivations of a user, and texts retrieved by a system or in the file of a system, or even in existence. Satisfaction, success, accomplishment, and the like are criteria for inferring motivational relevance."*

Having said that, Saracevic's definition of affective relevance has not had an unanimous acceptation by the community. Researchers such as Cosijn and Ingwersen [11] or Borlund [12] have argued that the concept does not refer to a specific type or kind of relevance but, instead, should be seen as an attribute of relevance that influences all other types of relevance. For instance, Cosijn and Ingwersen argue that affective relevance might act as an additional dimension, influencing other subjective relevance types (such as situational relevance, utility, etc.). Following Borlund's point of view, which is in the same direction, motivational or affective relevance is actually the cause for users to search. Therefore, according to her, intent, goals and motivations have to be seen as a characteristic of all the types of relevance: *"Thus, the 'drive' to want information is not an independent, specific type of relevance, but an inherent characteristic of relevance behavior in general"* [12].

Interestingly, Cosijn and Ingwersen [11] discuss the fact that affective and motivational relevance should be considered as two different types of relevance. In their view, motivational relevance is indeed defining the goals and motivations of the user, hence is seen as an independent characteristic influencing all the other kinds of relevance. Affective relevance, instead, is related to the affects and emotions a user experience when in an information-seeking task. Namely, they link their understanding of affective relevance with Barry's empirical investigation [17], and her types of relevance identified as *"criteria pertaining to the user's beliefs and preferences"*. Specifically, Cosijn and Ingwersen link affective relevance to Barry's *affectiveness*, that she defines as any kind of emotional responses to any aspect of a given document.

3   Physiological Computing in Information Retrieval

Physiological computing is a subfield of affective computing that has gained importance in the recent years. Researchers have studied how measuring psychophysiological measures can enhance current systems, by making the system aware of the user's cognitive, emotional and affective state [18–22]. Physiological computing goes far beyond information systems, and embraces a large range of fields, from human-computer interaction (HCI) and engineering to biofeedback and biocybernetic adaptation or human-robot interaction (HRI), to name a few [22]. Eye tracking, pupillometry, electroencephalography (EEG), cardiovascular measures (e.g. HRV, BPM), electro-dermal activity (EDA), and facial electromyography (fEMG) are some of the techniques used to infer user emotional and cognitive state [23]. Additionally, physiological computing systems involve real-time system adaptation to the user's psychophysiological measures. As opposed to other traditional implicit interaction techniques such as subliminal cueing [24, 25], physiological computing takes advantage of the ability of psychophysiological measures to implicitly indicate the users' cognitive, emotional and affective state in order to adapt accordingly, with the aim of a smooth, accurate and personalized interaction.

In the following sections we will discuss some of the previous work involving major psychophysiological measures, namely eye tracking and pupillometry, electroencephalography and other kind of peripheral physiology. The categorization is not obvious as one of the strengths of physiological computing systems is their ability to use multiple sensors and measures combined, in order to more accurately infer the user state. It is not our aim to exhaustively inspect the literature. Instead, we will focus on reporting few examples. Hopefully, the reader will get an insight on the multiple possibilities that psychophysiological input can bring to enhance human-machine symbiosis in information systems.

3.1   Neurophysiological Measures

The potential of brain imaging for enhancing information systems is outstanding. Every cognitive process takes place in the brain and the science of measuring and interpreting brain signals is ever growing. Two major techniques are being used for noninvasive brain imaging: electroencephalography (EEG) and functional magnetic resonance imaging (fMRI). In this section, we will focus on examples of how EEG measurements have been beneficial or are potentially applicable to information systems. We will exclude any example of research in the field of fMRI, as the technique is much more intrusive and expensive, therefore, less likely to become a viable input to everyday life information systems in the near future.

The field of Brain-Computer Interfaces (BCI) is a living example of how this field has raised the interest of both researchers and society [26]. When in physiological computing systems, the focus is on *passive BCIs*. In this specific case, the brain activity is passively measured in order to infer cognitive, affective or emotional states. The inferred state is then used as an input to the system. Therefore, interfaces where the user interacts by actively performing a mental task (i.e. *active BCIs*), are not considered. Electroencephalography has proven to be useful to implement *passive brain-computer interfaces*, for instance in order to measure task engagement (e.g. [27]), cognitive workload (e.g. [28]) or motivational intensity and fatigue (e.g. [29]). Many more examples follow, illustrating the high potential of using EEG to enhance information systems.

When in information retrieval, Eugster et al. have recently showed some hints on how it is potentially feasible, in a near future, to infer term relevance from brain signals only, by combining encephalography and machine learning techniques [30].

3.2   Eye-Derived Measures

Psychologists and cognitive scientists were the first to study eye-movements and to research eye tracking techniques. The first publication reporting an accurate eye-tacking system was back in 1901 by Dodge and Cline [31]. More than a century has passed and multiple eye tracking systems and applications have been studied and reported [32]. In this section we would like to focus on information science and information retrieval research. A large variety of eye-derived metrics have been used to infer the users' interests and their perceived relevance of information items. The goal of this section is to overview the systems that have used eye-movements analysis to enhance user modeling, adaptation and personalization.

Researchers such as Buscher have widely studied how to profit from eye movements in information retrieval settings [33–36]. We strongly encourage readers to go through one of his recent publications, where a large literature is covered, together with the results of two studies, indicating the possible benefits of taking into account eye movements and reading behaviors to improve information retrieval solutions [36]. Many more researchers have investigated the field. We would like to recall, for instance, recent work carried out by Ajanki et al. [37]. They studied the possibility of automatically generate implicit queries from eye movements. Their research fits very well in the field of physiological computing, as physiological responses (in this case eye-movements) were used in order to have an on-line adaptation of the system.

Many others have studied eye movements to infer relevance in documents, and a large variety of features have been used [38]. Fixation length (e.g. [39]), fixation count (e.g. [40]), thorough reading ratio (e.g. [36,41]), blink rate (e.g. [42]) or pupil size (e.g. [43,44]) are some examples.

3.3   Other Types of Peripheral Physiology

The analysis of psychophysiological data other than eye movements and brain activity is in the core of physiological computing. All sorts of measures are being used to infer user states, either using one simple measure or by combining multiple measures. For example, Kapoor et al. have shown how it is possible through different physiological channels (electro-dermal activity and pressure sensors, among others) to detect frustration [45]. Kataoka et al. showed how skin temperature can indicate stress [46]. Heart rate (HR) or blood volume pulse (BVP) are other examples of signals that have also been used as indicators of stress [47]. Facial electromyography (fEMG) has proven to be useful to infer the level of valence of the user, by measuring the activity on the muscles known as *zygomaticus major* (positive valence) and *corrugator supercilii* (negative valence) [48,49]. Electro-dermal activity is known as a robust measure to indicate level of arousal [50].

Regarding the special case of information retrieval, Arapakis has carried out research exploring the applicability of all kind of physiological measures to infer affective states during information seeking tasks, in order to enhance information retrieval systems [51–53]. The literature is broad, and multiple combinations of psychophysiological measures have been used to infer emotions and affective states. We refer the reader to Lopatovska and Arapakis recent paper on emotions in information science and information retrieval for a structured review of the recent advances in the field [54].

More "superficial" measures such as video-based recognition of facial expressions have also been analyzed in information systems in order to infer users' emotions (e.g. [55]). In this section we have not covered face and voice recognition or gestures tracking for one main reason. In the present research, towards an enhanced symbiotic interaction, we are especially interested in the underlying physiological responses of a given affective or motivational state. The psychophysiological responses described in this section are strongly related with the user's state while being non-voluntary and, therefore, less likely to be faked.

4   Discussion

The concept of relevance is not strongly defined. Different authors show different points of view, and the information science community has not adopted a unified notion of relevance yet. Mizzaro is one of the authors that have expressed their concerns regarding the inconsistency in the terminology used in the relevance literature, as terms are given different meanings by different authors and authors use terms in an ambiguous manner or use different terms as synonyms [9]. Nevertheless, common views are shared among the researchers, especially regarding the complex, multifaceted and dynamic nature of relevance. Even though the literature on relevance includes a much broader range of concepts than the described in this work (e.g. utility, situational relevance), this paper aims to raise a discussion around some very specific facets of relevance. As with relevance in general, no strict terminology exists around psychological, affective and motivational relevance, and their definitions commonly overlap. For instance, Saracevic defines cognitive relevance or pertinence, which is a very close concept to Harter's psychological relevance [4], and according to different authors, motivational and affective relevance are described as a unique or as two different concepts.

The present paper overviews some uses of psychophysiological signals in information science and information retrieval, that have been used to infer users' perception of relevance and interest in information items. Eye tracking research and term-relevance prediction through encephalography are some of the overviewed methods. It has to be noted that it is not our aim to replace these methods. Instead, we think it might be useful to investigate further methods that are able to work in a complementary way, in order to enhance user experience, system performance and, ultimately, human-machine symbiosis. The line of research we propose is derived from the definition of psychological, affective and motivational relevance. These definitions agree in the fact that users' cognitive context and cognitive, affective and motivational state play a role in the perception of relevance that users have on information items. Therefore, we would like to use the different theoretical frameworks as an inspiration to base our approach, in which we believe that measuring and modeling user cognitive and affective state might be of great use to enhance current personalized information retrieval systems. In this way, we aim to reach a much more informed understanding of the users' perception of relevance, as it is a main factor in the user's subjective formulation of relevance. That is, the system will be enhanced with knowledge about the user cognitive, affective motivational state. As previously discussed, and in light of the broad literature regarding physiological computing, we believe that those methods are perfectly suitable to be used in the field of personalized information retrieval systems. Nevertheless, when designing such symbiotic systems, it is worthwhile to keep in mind that the current techniques to infer user states through psychophysiological measures are not faultless, and that users' acceptance of physiological recording devices is still being studied [56].

Below we formulate a set of research questions and, for each of them, we propose a method to bring an answer to them. In this way, we frame an approach to the exciting task of merging theoretical frameworks in information science and state-of-the-art physiological computing systems to enhance adaptive information retrieval systems.

Q1. Which physiological measures and which cognitive states are best suited to indicate affective relevance? It is needed to thoroughly review the literature on physiological measures used to infer cognitive states in order to comprehensively identify which are the cognitive states that might potentially influence perception of relevance, and which are the physiological measures that are able to indicate them. The literature in human-computer interaction regarding the use of psychophysiological measures is broad, and it is important to identify which are the studies that are potentially applicable to information-seeking and information retrieval systems. It is important to note that a large amount of these studies are carried out in very specific and controlled experimental conditions, and might not be applicable to our proposed approach.

Q2. Which are the information-seeking scenarios where the modeling of affective relevance is more pertinent in order to enhance current information retrieval solutions? We hypothesize that complex search tasks are best suited for taking advantage of psychological, motivational and affective relevance metrics. Search tasks with a large exploratory component might benefit more than simple lookup tasks, where the cognitive effort and context is relatively low. Research should be carried out through user studies, in order to measure different cognitive states such as level of frustration, motivation, engagement, confusion, etc. and their relationship with perceived relevance in different search scenarios. In this way, search-tasks where the modeling of affective psychological and motivational relevance is more pertinent will be identified.

Q3. How to model affective relevance? Once the information-search tasks and the cognitive states have been identified, it is important to define, implement and test several user models based on affective and cognitive states of relevance in order to discuss how these models can be applied to currently available information retrieval systems. A possible approach are latent variable models, based on the different cognitive and affective states.

Q4. To what extend modeling and implementing affective relevance in an information-seeking system enhances current information retrieval solutions? Once the scenarios and measures have been identified, and the models have been implemented, the different components need to be merged together into a real information retrieval system that makes use of the affective relevance model through psychophysiological measurements. The adaptation and personalization mechanisms need to be defined and tested before it is possible to evaluate the affective relevance-based system in comparison with current I.R systems. It is worth to consider as well mixt systems where the affective relevance model coexists with other personalization techniques.

The above-presented research questions are intended to be indicative, but are not exhaustive. The aim is to raise a discussion around the possibilities that new physiological computing techniques bring to the field of information retrieval. These techniques bring the opportunity to implement concepts that have remained in the theoretical frameworks of the information science literature for decades. We hope that this paper, by presenting an overview on both the theory of relevance and state-of-the-art physiological computing techniques, will engage the community in a fruitful discussion around the proposed challenges and research questions.

References

1. Schamber, L.: Relevance and information behavior. Ann. Rev. Inf. Sci. Technol. (ARIST) **29**, 3–48 (1994)
2. Jacucci, G., Spagnolli, A., Freeman, J., Gamberini, L.: Symbiotic interaction: a critical definition and comparison to other human-computer paradigms. In: Jacucci, G., Gamberini, L., Freeman, J., Spagnolli, A. (eds.) Symbiotic 2014. LNCS, vol. 8820, pp. 3–20. Springer, Heidelberg (2014)
3. Saracevic, T.: Relevance: a review of and a framework for the thinking on the notion in information science. J. Am. Soc. Inform. Sci. **26**(6), 321–343 (1975)
4. Saracevic, T.: Relevance reconsidered. In: Proceedings of the Second Conference on Conceptions of Library and Information Science (CoLIS 2), pp. 201–218. ACM Press (1996)
5. Saracevic, T.: The stratified model of information retrieval interaction: extension and applications. In: Proceedings of the Annual Meeting-American Society for Information Science. vol. 34, pp. 313–327. Learned Information (Europe) LTD (1997)

6. Saracevic, T.: Relevance: a review of the literature and a framework for thinking on the notion in information science. Part II: Nature and manifestations of relevance. J. Am. Soc. Inform. Sci. Technol. **58**(13), 1915–1933 (2007)
7. Saracevic, T.: Relevance: a review of the literature and a framework for thinking on the notion in information science. Part III: Behavior and effects of relevance. J. Am. Soc. Inform. Sci. Technol. **58**(13), 2126–2144 (2007)
8. Mizzaro, S.: Relevance: the whole history. J. Am. Soc. Inform. Sci. **48**(9), 810–832 (1997)
9. Mizzaro, S.: How many relevances in information retrieval? Interact. Comput. **10**(3), 303–320 (1998)
10. Ingwersen, P.: Cognitive perspectives of information retrieval interaction: elements of a cognitive ir theory. J. Doc. **52**(1), 3–50 (1996)
11. Cosijn, E., Ingwersen, P.: Dimensions of relevance. Inf. Process. Manage. **36**(4), 533–550 (2000)
12. Borlund, P.: The concept of relevance in IR. J. Am. Soc. Inform. Sci. Technol. **54**(10), 913–925 (2003)
13. Borlund, P., Ingwersen, P.: Measures of relative relevance and ranked half-life: performance indicators for interactive IR. In: Proceedings of the 21st Annual International ACM SIGIR Conference on Research and Development in Information Retrieval, pp. 324–331. ACM (1998)
14. Sperber, D., Wilson, D.: Relevance: Communication and Cognition. Harvard University Press, Cambridge (1986)
15. Harter, S.P.: Psychological relevance and information science. J. Am. Soc. Inf. Sci. **43**(9), 602–615 (1992)
16. Schutz, A., Zaner, R.: Reflections on the Problem of Relevance. Greenwood Press, Westport (1970)
17. Barry, C.L.: User-defined relevance criteria: an exploratory study. JASIS **45**(3), 149–159 (1994)
18. Picard, R.W., Vyzas, E., Healey, J.: Toward machine emotional intelligence: analysis of affective physiological state. IEEE Trans. Pattern Anal. Mach. Intell. **23**(10), 1175–1191 (2001)
19. Allanson, J., Wilson, G.M.: Physiological computing. In: CHI'02 Extended Abstracts on Human Factors in Computing Systems, pp. 912–913. ACM (2002)
20. Allanson, J., Fairclough, S.H.: A research agenda for physiological computing. Interact. Comput. **16**(5), 857–878 (2004)
21. Fairclough, S.H.: Fundamentals of physiological computing. Interact. Comput. **21**(1), 133–145 (2009)
22. Fairclough, S.H., Gilleade, K.: Advances in Physiological Computing. Springer, London (2014)
23. Cacioppo, J.T., Tassinary, L.G., Berntson, G.G., et al.: Handbook of Psychophysiology, vol. 2. Cambridge University Press, New York (2007)
24. Negri, P., Gamberini, L., Cutini, S.: A review of researches on subliminal techniques for implicit interaction in symbiotic systems. In: Jacucci, G., Gamberini, L., Freeman, J., Spagnolli, A. (eds.) Symbiotic 2014. LNCS, vol. 8820, pp. 47–60. Springer, Heidelberg (2014)
25. Barral, O., Aranyi, G., Kouider, S., Lindsay, A., Prins, H., Ahmed, I., Jacucci, G., Negri, P., Gamberini, L., Pizzi, D., Cavazza, M.: Covert persuasive technologies: bringing subliminal cues to Human-Computer interaction. In: Spagnolli, A., Chittaro, L., Gamberini, L. (eds.) PERSUASIVE 2014. LNCS, vol. 8462, pp. 1–12. Springer, Heidelberg (2014)

26. Müller, K.R., Kübler, A.: An introduction to brain-computer. In: Dornhege, G., Millan, J.R., Hinterberger, T., McFarland, D.J., Muller, K.R. (eds.) Toward Brain Computer Interfacing, pp. 1–25. Massachusetts Institute of Technology Press, Cambridge (2007)

27. Fairclough, S.H., Ewing, K.C., Roberts, J.: Measuring task engagement as an input to physiological computing. In: 3rd International Conference on Affective Computing and Intelligent Interaction and Workshops, 2009, ACII 2009, pp. 1–9. IEEE (2009)

28. Gevins, A., Smith, M.E.: Neurophysiological measures of cognitive workload during human-computer interaction. Theor. Issues Ergonomics Sci. 4(1–2), 113–131 (2003)

29. Lorist, M.M., Bezdan, E., ten Caat, M., Span, M.M., Roerdink, J.B., Maurits, N.M.: The influence of mental fatigue and motivation on neural network dynamics; an EEG coherence study. Brain Res. 1270, 95–106 (2009)

30. Eugster, M.J., Ruotsalo, T., Spapé, M.M., Kosunen, I., Barral, O., Ravaja, N., Jacucci, G., Kaski, S.: Predicting term-relevance from brain signals. In: Proceedings of the 37th International ACM SIGIR Conference on Research and Development in Information Retrieval, SIGIR '14, pp. 425–434. ACM, New York (2014)

31. Dodge, R., Cline, T.S.: The angle velocity of eye movements. Psychol. Rev. 8(2), 145 (1901)

32. Duchowski, A.: Eye Tracking Methodology: Theory and Practice, vol. 373. Springer, London (2007)

33. Buscher, G., Dengel, A., van Elst, L.: Eye movements as implicit relevance feedback. In: CHI'08 Extended Abstracts on Human Factors in Computing Systems, pp. 2991–2996. ACM (2008)

34. Buscher, G., Dengel, A., van Elst, L.: Query expansion using gaze-based feedback on the subdocument level. In: Proceedings of the 31st Annual International ACM SIGIR Conference on Research and Development in Information Retrieval, pp. 387–394. ACM (2008)

35. Buscher, G., Cutrell, E., Morris, M.R.: What do you see when you're surfing?: using eye tracking to predict salient regions of web pages. In: Proceedings of the SIGCHI Conference on Human Factors in Computing Systems, pp. 21–30. ACM (2009)

36. Buscher, G., Dengel, A., Biedert, R., Elst, L.V.: Attentive documents: eye tracking as implicit feedback for information retrieval and beyond. ACM Trans. Interact. Intell. Syst. (TiiS) 1(2), 9 (2012)

37. Ajanki, A., Hardoon, D.R., Kaski, S., Puolamäki, K., Shawe-Taylor, J.: Can eyes reveal interest? implicit queries from gaze patterns. User Model. User-Adap. Inter. 19(4), 307–339 (2009)

38. Salojärvi, J., Puolamäki, K., Simola, J., Kovanen, L., Kojo, I., Kaski, S.: Inferring relevance from eye movements: feature extraction (2005)

39. Loboda, T.D., Brusilovsky, P., Brunstein, J.: Inferring word relevance from eye-movements of readers. In: Proceedings of the 16th International Conference on Intelligent User Interfaces, pp. 175–184. ACM (2011)

40. Puolamäki, K., Salojärvi, J., Savia, E., Simola, J., Kaski, S.: Combining eye movements and collaborative filtering for proactive information retrieval. In: Proceedings of the 28th Annual International ACM SIGIR Conference on Research and Development in Information Retrieval, pp. 146–153. ACM (2005)

41. Moe, K.K., Jensen, J.M., Larsen, B.: A qualitative look at eye-tracking for implicit relevance feedback. In: Proceedings of the Workshop on Context-Based Information Retrieval, vol. 326, pp. 36–47. Citeseer (2007)

42. Goldberg, J.H., Wichansky, A.M.: Eye tracking in usability evaluation: a practitioner's guide. In: Hyona, J., Radach, R., Deubel, H. (eds.) The Mind's Eye: Cognitive and Applied Aspects of Eye Movement Research, pp. 573–605. Elsevier, Amsterdam (2003)

43. Oliveira, F.T., Aula, A., Russell, D.M.: Discriminating the relevance of web search results with measures of pupil size. In: Proceedings of the SIGCHI Conference on Human Factors in Computing Systems, pp. 2209–2212. ACM (2009)

44. Barral, O., Kosunen, I., Jacucci, G.: Influence of reading speed on pupil size as a measure of perceived relevance. In: Proceedings of the Joint Workshop on Personalized Information Access (PIA 2014), in Conjunction with the 22nd Conference on User Modeling, Adaptation and Personalization (UMAP 2014)

45. Kapoor, A., Burleson, W., Picard, R.W.: Automatic prediction of frustration. Int. J. Hum. Comput. Stud. **65**(8), 724–736 (2007)

46. Kataoka, H., Kano, H., Yoshida, H., Saijo, A., Yasuda, M., Osumi, M.: Development of a skin temperature measuring system for non-contact stress evaluation. In: Engineering in Medicine and Biology Society, 1998, Proceedings of the 20th Annual International Conference of the IEEE, vol. 2, pp. 940–943. IEEE (1998)

47. Wilson, G.M., Sasse, M.A.: Listen to your heart rate: counting the cost of media quality. In: Paiva, A.C.R. (ed.) IWAI 1999. LNCS, vol. 1814, pp. 9–20. Springer, Heidelberg (2000)

48. Partala, T.: Affective Information in Human-Computer Interaction. Tampereen Yliopisto, Tampere (2005)

49. Gilroy, S., Porteous, J., Charles, F., Cavazza, M.: Exploring passive user interaction for adaptive narratives. In: Proceedings of the 2012 ACM International Conference on Intelligent User Interfaces, pp. 119–128. ACM (2012)

50. Boucsein, W.: Electrodermal Activity. Springer, New York (2012)

51. Arapakis, I., Jose, J.M., Gray, P.D.: Affective feedback: an investigation into the role of emotions in the information seeking process. In: Proceedings of the 31st Annual International ACM SIGIR Conference on Research and Development in Information Retrieval, pp. 395–402. ACM (2008)

52. Arapakis, I., Konstas, I., Jose, J.M.: Using facial expressions and peripheral physiological signals as implicit indicators of topical relevance. In: Proceedings of the 17th ACM International Conference on Multimedia, pp. 461–470. ACM (2009)

53. Arapakis, I., Athanasakos, K., Jose, J.M.: A comparison of general vs personalised affective models for the prediction of topical relevance. In: Proceedings of the 33rd International ACM SIGIR Conference on Research and Development in Information Retrieval, pp. 371–378. ACM (2010)

54. Lopatovska, I., Arapakis, I.: Theories, methods and current research on emotions in library and information science, information retrieval and human-computer interaction. Inf. Process. Manage. **47**(4), 575–592 (2011)

55. Arapakis, I., Moshfeghi, Y., Joho, H., Ren, R., Hannah, D., Jose, J.M.: Enriching user profiling with affective features for the improvement of a multimodal recommender system. In: Proceedings of the ACM International Conference on Image and Video Retrieval, p. 29. ACM (2009)

56. Spagnolli, A., Guardigli, E., Orso, V., Varotto, A., Gamberini, L.: Measuring user acceptance of wearable symbiotic devices: validation study across application scenarios. In: Jacucci, G., Gamberini, L., Freeman, J., Spagnolli, A. (eds.) Symbiotic 2014. LNCS, vol. 8820, pp. 87–98. Springer, Heidelberg (2014)

A Review of the Research on Subliminal Techniques for Implicit Interaction in Symbiotic Systems

Paolo Negri[1(✉)], Luciano Gamberini[1,2], and Simone Cutini[1]

[1] Department of General Psychology,
University of Padua, Padua, Italy
paolonegri0@gmail.com,
{luciano.gamberini,simone.cutini}@unipd.it
[2] Human Inspired Technology Research Centre,
University of Padua, Padua, Italy
http://htlab.psy.unipd.it/

Abstract. Subliminal perception is a long-standing topic in psychology, which has been strongly debated throughout the twentieth century. Recently, unconscious information processing has gained attention in human-computer interaction (HCI) research on the basis that subliminal stimulation can covertly trigger automatic responses without generating mental workload. The aim is to increase the interaction efficiency between humans and systems by embedding subliminal stimuli in user interfaces. Moreover, the currently thriving research on adaptive and symbiotic systems makes the interest for unconscious processes even greater.

The purpose of the present paper is to give an overview of both the most recent findings about subliminal stimuli applied to concrete contexts and the main stimulation techniques to obtain unconscious perception. The techniques reviewed here are the binocular rivalry, visual masking, visual crowding, and rapid serial visual presentation with some latest variants of these classic paradigms.

Keywords: Symbiotic and adaptive system · Subliminal perception

1 Introduction

A visual stimulus is defined as subliminal when, due to particular features (e.g., a very brief presentation), it cannot be consciously perceived. Despite the lack of awareness about its existence, the subliminal stimulus can affect the human's mental activity and behavior.

Research on subliminal perception has been one of the most debated topics in psychology over the twentieth century (see [1] for a review). Scientists have adopted ever more sophisticated and strictly controlled subliminal techniques and methods of subliminal stimulation that have often led them to draw different conclusions about the existence of non-conscious processes [1]. There is a variety of conditions under which the visual perception can be systematically manipulated to obtain unconscious perception, and each subliminal technique capitalizes on one or more of these conditions

© Springer International Publishing Switzerland 2014
G. Jacucci et al. (Eds.): Symbiotic 2014, LNCS 8820, pp. 47–58, 2014.
DOI: 10.1007/978-3-319-13500-7_4

(see Sect. 3). Nowadays, while there are no more doubts about the existence of a perception without awareness, unconscious information processing has gained new attention in HCI research. By embedding subliminal stimuli in user interfaces, the goal is to increase the interaction efficiency between humans and systems. Such an idea exploits the fact that subliminal information processing is not worsened when it occurs in parallel with other conscious cognitive processes [2]. Therefore, the integration of subliminal stimuli in user interfaces could enrich the information transmission between a computer and a human in an effective way when a large amount of data needs to be processed, and the user's cognitive system is at risk of becoming overloaded. Furthermore, the currently thriving research on adaptive and symbiotic systems [3–7] makes the interest in unconscious processes even greater. Symbiotic relationships between humans and computers need a reciprocal and deeper understanding, which could be achieved by expanding the bandwidth of the ordinary communication with machines beyond the mere symbolic exchange. On the one hand, the computer can help to achieve a deeper understanding of the human state by capturing implicit behaviors that might be recognized with sensing technologies or other implicit signals. On the other hand, by means of subliminal stimulation, the computer can help to covertly trigger automatic psychological responses in the user, and this is the reason that subliminal stimuli would represent a fundamental branch of this reciprocal deeper interaction. Subliminal stimuli can ignite unconscious mental activity, allowing a deepening of user-system interaction below the "limen" of human consciousness.

As a good illustration of the symbiotic relationship between humans and computers, an interesting conception of a synthetic reality platform useful to support 3D visualization of realistic information has recently been proposed [3]. This platform, endowed with an intelligent narrative engine, has been conceived to investigate the use of subliminal cues and implicit responses to induce specific user experiences during interactions with 3D models. The system uses both explicit interaction and implicit data (psychophysiology) related to the real-time user experience aiming to adapt itself to the interaction and to assist the user exploration. In this context, subliminal stimuli are conceived as a mechanism to covertly guide the user during his exploratory experience. Goals for this covert guide might be (a) to assist the user during critical phases, (b) to manage the exploration in order to shift his interest to other dimensions to be noticed, and (c) to maintain the user's interest focused on a given dimension [3]. The user's experience can benefit from this hidden assistance because subliminal stimulation does not generate mental workload [8, 9] and does not interrupt his task at hand.

As a counterpart of these advantages, information perceived without awareness ignites implicit cognitive processes that, precisely because they are implicit, cannot be controlled by the perceiver [10]. In other words, processing with and without awareness leads to qualitatively different results: awareness enables the intentional use of information, whereas unconscious processing only increases the likelihood that information will be used in a subsequent task [2]. This does not mean that the human under subliminal stimulation is forced to act like an automaton; rather, it simply means that subliminal perception does not give rise to a symbolic representation that is accessible to critical thinking.

With respect to the "intentionality of use", it should be noted that slightly different positions and theoretical explanations exist. This article is not intended to enter into a detailed dissertation on this delicate subject; however, it is worth reporting that some authors [4, 11] suggest that volitional behavior and automatic responses do not differ regarding the level of intentional control. Instead, they suggest that, in the former, the information is consciously accessible as propositional, and in the latter, it is accessible as a "feeling of." We mean that even subliminal stimulation results in a mental product that is available for intentional control: qualia-like states, feelings that the individual may or may not be inclined to act on [4].

However, access to consciousness seems to be mandatory for crucial mental computations, such as durable information maintenance and planning novel strategy [12]. Indeed, several mental operations cannot be accomplished unconsciously, and research on the use of subliminal stimuli in user interfaces should consider both their advantages and limitations.

To date, very few efforts have been made to address some crucial issues concerning an effective use of subliminal stimuli in applied concrete scenarios. Indeed, although subliminal perception is a long-standing topic in psychology, the large amount of research demonstrates that it is mainly used in abstract tasks and contexts aiming at investigating visual perception and its neural correlates. Conversely, more ecological scenarios are needed, and it is essential to systematically investigate what types of stimuli are more effective (e.g., symbols, words) in these scenarios, for what types of tasks (e.g., item selection, learning, attention grabbing) and, relevant for the present work, with which type of subliminal technique.

2 Related Work

In this section, we report a few representative studies that broadly summarize the main properties of subliminal stimulation applied to realistic contexts.

The first study in our report [9], which uses visual masking,[1] demonstrated the effectiveness of subliminal cueing[2] as a memory retrieval aid, using cues delivered in a head-mounted display. In another study [13], visual masking was applied in the context of a virtual tutoring system. The results showed that subliminal cueing elicited better performances as well as better affective states throughout the lesson. Other studies [14, 15] have focused on the influence of subliminal cueing on visual search tasks in graphical user interfaces (GUIs).

More recent studies embedded subliminal interaction techniques in 3D virtual environments and mixed-reality systems and strictly controlled the visibility of stimuli, which is mandatory in order to assert a genuine subliminal effect.

In the first of these studies [5], the use of subliminal stimuli has been considered to covertly bias the selection behavior among objects in a 3D virtual environment. The participants' task was to select one food item out of two options in a 3D model of a refrigerator, pick it up, and place it on an adjacent table. Before the selection behavior, one of the foods was subliminally cued with the visual masking technique. The results

---

[1] Visual masking is one of the techniques used to make a visual stimulus subliminal; this technique is addressed in the section titled "Unconscious Perception."

[2] Subliminal cueing consists of the bias affecting choice among alternative targets or actions caused by a preceding subliminal stimulus.

showed that cued items were selected significantly more than non-cued ones. This happened especially when participants responded within 1 s after the disappearance of the subliminal cue. In one condition, the subliminal effect was more long-lasting: when the stimulation consisted in a rapid sequence (three times) of the exposure of the cue.

Overall, these results lead to some important considerations: first of all, it has been demonstrated that subliminal stimuli can be used to bias participants' responses in a concrete selection task among 3D virtual objects. However, it also emerged that the effect of subliminal stimulation is extremely short-lived and even quite weak. The designing of user interfaces embedding subliminal stimuli should appropriately take into account these findings.

Another study [4] investigated the impact of subliminal cues in an immersive navigation task. Participants were seated in an immersive mixed-reality system, and their task was to navigate within a maze. During the navigation, participants had to repeatedly make dichotomous choices between two alternative paths, and prior to each choice, they were subliminally exposed (using visual masking) either to an aversive (a spider) or to a neutral stimulus. Therefore, some maze paths were labeled with the aversive stimulus, and others were labeled with the neutral one. The results showed that those paths that had been negatively labeled (with the spider) were more likely to be avoided. The main finding of this experiment consists of the demonstration that arousing subliminal stimuli can bias the decision-making during navigation tasks in an immersive mixed-reality system. However, despite the relevance of these results, as emerged in the abovementioned work [5], the magnitude of the observed effect was comparable with that reported in the related literature but quite weak to effectively support realistic applications.

Subliminal interaction techniques will turn out to be powerful tools in HCI and more specifically in symbiotic systems, provided that the design of these systems takes into account some hallmarks of unconscious information processing. To briefly sum up, subliminal stimulation seems to be suitable for a variety of uses in concrete scenarios as those of memory retrieval [9] and problem solving [13] to bias decision-making in selection behavior [5, 16] and navigation tasks [4]. In the specific context of symbiotic systems, such potential can be exploited to assist the user during critical phases or to manage his interaction without interrupting the task at hand [3] or generating mental workload [8, 9]. To achieve these results, it is mandatory to remember that some crucial mental computations are not unconsciously executable [12] and that subliminal perception gives rise to low-confidence knowledge that users can manage, provided that they are willing to base their actions on qualia-like thoughts [4].

However, the main issues concerning subliminal stimulation are its weak and short-lived effect. Furthermore, as emerged from the reviewed literature, studies predominantly used the visual masking technique to make the experimental stimuli subliminal, and this method requires a stimulation to be delivered foveally, just to the center of the visual field. This circumstance obviously interferes with the execution of the task at hand.

To overcome these latter issues, a multiplicity of other techniques could be considered. There is a variety of situations and conditions under which the visual perception of a normal-sighted person can be systematically manipulated to obtain an unconscious perception. At least under some conditions, different techniques seemed to be more powerful than visual masking [17], and to the best of our knowledge, they have never been used so far in an HCI context.

The main aim of the following section is to give an overview of some of the main techniques that can be used to make a visual stimulus unconscious. The techniques here described are the binocular rivalry, visual masking, and their variants, followed by visual crowding and rapid serial visual presentation.

3   Unconscious Perception

3.1   Binocular Rivalry

Binocular rivalry is the most striking variant of perceptual behavior called bistability: a continuous oscillation of conscious perception between two alternative percepts that we may appreciate by observing the well-known ambiguous figures (e.g., the Necker cube; [18]). In ambiguous figures and in "monocular rivalry" [19], only one stimulus is presented to both eyes, resulting in a "multistable perception" [20]. This multistable perception is a phenomenon consisting of an unpredictable fluctuation of the consciousness between two different percepts. The two percepts are evoked by the same stimulus, which, due to its ambiguity, does not allow a single interpretation. In binocular rivalry, the ambiguity is specifically created through a dichoptic vision (i.e., left and right eyes are presented with a slightly different image). Therefore, binocular rivalry is a mechanism that allows the brain to address the unnatural condition in which the same position in the space is occupied by different objects. If this is the case, a breakdown of ordinary perception occurs, forcing the brain to deal with this anomaly (physical impossibility) without merging the information coming from the two retinas. As a result, these representations become alternately dominant in the consciousness [21].

A remarkable feature of rivalry is the unpredictability of its timing. It seems impossible to forecast the length of the period during which the percept related to one eye will be consciously perceived or when the percept will be suppressed and the access to visual awareness will be transferred to the percept coming from the other eye. Several studies have been attempted to establish whether the variability of the durations is somehow a deterministic dimension or whether it is fully stochastic: results seem to favor the latter hypothesis [22].

Variants of binocular rivalry. Many findings about binocular rivalry have been capitalized in the conception of sophisticated methods to create rivalry conditions, which allow researchers to exert a high degree of control over the timing of the perceptual alternation. The first of these methods is the "binocular rivalry flash suppression" – BRFS – often simply called "flash suppression" [23]. Flash suppression consists of the presentation of two dissimilar visual targets in dichoptical vision, but unlike the simple binocular rivalry, the two visual patterns are not presented simultaneously. Rather, they are presented in an asynchronous way. That is, the first stimulus is presented monocularly and, in the absence of any blank interval, is followed by the flashed presentation of a different pattern in the contralateral eye, whereas the stimulus remains the same ipsilaterally. Flashing a different contralateral stimulus, the first stimulus presented (monocularly) disappears from consciousness; notably, its suppression can persist for several seconds.

Similar to BRFS, the "generalized flash suppression" – GFS – [24] consists of the monoptical presentation of a salient stimulus (as a luminous dot) followed (after several hundred ms) by the dichoptic flashed presentation of a surround filled in with moving or stationary dots. After the onset of the surround, the target abruptly disappears from consciousness. Interestingly, unlike the other rivalries, the present one does not present interocular discrepancy between different stimuli. Indeed, the disappearing dot is not coupled with another different, spatially overlapped stimulus. GFS combines some features of "flash suppression" with elements of "motion-induced blindness" [25], consisting of a moving surround, even if the effect can take place even with a stationary surround [24].

Definitely, the "continuous flash suppression" – CFS – [26] is the strongest type of binocular rivalry, and it leads to a longer suppression effect and to greater control over the perceptual alternation. This paradigm combines binocular rivalry and flash suppression, and it requires the presentation of a permanent stationary pattern to an eye and an ongoing flashing of different images into the other eye (e.g., mondrians). The result is the disappearance from consciousness of the stationary pattern, which lasts for a few minutes.

3.2 Visual Masking

The visual masking is the reduction of the visibility of a stimulus due to the close in time and space appearance of another visual stimulus. In the visual masking, there is a target visual object (e.g., a shape) that is briefly presented (for a few milliseconds) and immediately followed (or preceded) by a mask (e.g., another shape): crucially, the mask appearance makes the target inaccessible to the individual's consciousness.

There are several types of visual masking, resulting from different combinations of targets, masks, and timings. One of the main temporal parameters to be taken into account is the stimulus onset asynchrony (SOA), which is the delay between the onset of the target and that of the mask. A positive SOA means that a mask follows the target, generating a backward masking or metacontrast (if stimuli are as in 1a); with negative SOA, the mask precedes the target, producing forward masking or paracontrast (if stimuli are as in 1a) [27].

Regarding the types of masks, besides the earliest masking by light [28], consisting of the presentation of an abrupt increase or decrease of light, different targets and masks have been used in masking experiments. Figure 1 depicts some examples that generate

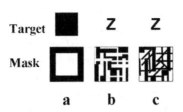

Fig. 1. Typical stimuli of: (a) paracontrast and metacontrast, (b) masking by noise, and (c) masking by structure (adapted from [27])

three different types of masking. When the contours of the two stimuli do not overlap but are close or contiguous (Fig. 1a), the masking is defined as either paracontrast or metacontrast (depending on whether the mask appears before or after the target, respectively). When the contours of the two stimuli overlap and the mask is a random structure that does not share structural features with the target (Fig. 1b), the masking is defined as masking by noise. Finally, when the contours of the stimuli overlap and the mask shares some structural features with the target (Fig. 1c), the masking is identified as masking by structure.

Perhaps the most striking example of visual masking for non-overlapped stimuli is the "Standing Wave of Invisibility" [29]. This paradigm combines forward and backward masking into a sequence of targets and masks separated by short intervals. By properly adjusting the time parameters of presentation (SOA, inter-stimulus interval [ISI], and stimulus termination asynchrony [STA]), a veil of invisibility covers the targets, which do not reach consciousness because of the inhibitory impact of the preceding and following masks.

3.3   Visual Crowding

The first studies on crowding date back to the early 20th century. In 1923, Korte described a visual crowding condition akin to the one in Fig. 2 as follows: "It is as if there is a pressure on both sides of the word that tends to compress it. Then the stronger, i.e. the more salient or dominant letters, are preserved and they 'squash' the weaker, i.e. the less salient letters, between them" [30].

When foveating the cross in Fig. 2, the crowding effect in the individual's peripheral vision can be experienced by moving from the left square, in which there is a single target letter (which is easily recognizable), to the right one, in which the target and several flankers are cluttered and thereby perceived in a jumbled way. Thus, as is generally defined, crowding is "the deleterious influence of nearby contours on visual discrimination" [31] that is, a progressive decrease in the ability to recognize objects in clutters. Crowding is ubiquitous in the visual field, although the higher the degree of eccentricity, the more dramatic is the effect.

Intuitively, this visual phenomenon is a pervasive presence in our lives, and this can explain the large amount of studies about crowding in a wide range of experimental ways, as in letter recognition tasks [32], face recognition tasks [33], and many more [31].

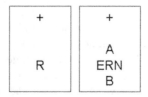

Fig. 2. The reader can experience the visual crowding of letters by fixing the gaze on the cross (adapted from [31])

A simple and important rule recently proposed [34] is the Bouma's rule, which states that there is a "critical spacing" for crowding [35]. The minimum distance between the target and the flankers which still allows the individual to distinctly perceive the target as an isolated element is proportional to the target eccentricity. Thus, the greater the degree of eccentricity, the greater must be the distance between stimuli to avoid a "crowded" perception.

Recently, a new approach called gaze-contingent crowding (GCC) has been proposed [36]. This approach seems to be particularly suitable for use in realistic scenarios because it consists of the long-lasting presentation of a crowded target stimulus (constantly unavailable to conscious perception) located in the periphery of the visual field. The foveal access of the target is prevented by means of the substitution of the target stimulus once an eye-tracking system has detected the movement of the eye gaze from a specific fixation point.

In comparison to the most widely used visual masking, the GCC leaves free the central part of the visual field, allowing the user to explore the visual information in a natural way.

3.4   Rapid Serial Visual Presentation

An emblematic example of attentional failure occurring when the human cognitive system is pushed to its limits is the attentional blink. In rapid serial visual presentation (RSVP) paradigms (Fig. 3), participants have to detect a target letter (or "T1") and a subsequent probe letter (or "T2"). If T2 appears within a time window ranging between 200 and 400/500 ms after the T1 onset, the correct identification of T1 markedly impairs the detection of T2. This temporary inability to detect stimuli has been named attentional blink [37], and it can be explained by three classes of theoretical accounts. The first class taps on an "active suppression" of the post-T (i.e., the first stimulus following T1), as is the case of the "attentional suppression model" [37].

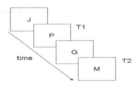

Fig. 3. Illustration of a rapid serial visual presentation used to observe the attentional blink

This model proposes that the cognitive system implements an active suppression of the post-T in order to prevent an erroneous perceptual fusion of sensory elements belonging to different objects. That is, because the target is immediately followed by another letter (in Fig. 3, letter "G") and because the time needed to identify the target exceeds the onset-to-onset time between the target and the next stimulus ("G"), the cognitive system suppresses any new sensory input, even T2. Consequently, the effect of the attentional blink consists of preventing the identification of further targets by shutting down attentional resources.

A second class of explanations involves capacity limitations. For example, the two-stage model [38, 39] assumes that a detection process consists of a short-lived early rapid detection that triggers a following "capacity-limited" stage of the identification process. During the course of the "capacity-limited" stage, other rapid detection stages may occur, although no further "capacity-limited" processes can be activated. Consequently, because of its short life, the rapid detection of a second target could vanish before the storage of the first target is completed, thus preventing its consolidation in the capacity limited stage.

The third class of theories refers to the failed retrieval of T2 from working memory, rather than a mere attentional limit [40]. It has been found [41] that the words presented within the time window of the attentional blink gave rise to the N400 peak. N400 is a negative electrophysiological peak occurring at 400 ms post-stimulus, which is related to the mismatch between a linguistic stimulus and its semantic context. Therefore, the presence of the N400 peak during attentional blink demonstrates that words are semantically processed even if they cannot reach awareness, providing support to the account referring to an impaired memory retrieval.

To date, in HCI research, the RSVPs have been studied as tools to support rapid information browsing based on the ability of the human visual system to recognize a known image in a pre-attentive way [42]. In this context, research has been focused only on which features the RSVPs should have to avoid the attentional blink. However, it seems that some level of semantic processing takes place for the stimulus presented during the blink [40], and one might wonder whether it is possible to somehow take advantage of this unconscious processing.

4 Conclusion

The currently thriving research on adaptive and symbiotic systems in HCI has awakened the interest of researchers in unconscious information processing. In these contexts, the aim is to establish between human and system an implicit communication loop, and the subliminal stimulation could represent a fundamental branch of this reciprocal hidden interaction for a variety of reasons, provided that the design of these systems takes into account subliminal stimulation's strengths and weaknesses. Subliminal stimulation seems to be suitable for a variety of uses in concrete scenarios. These potentials can be exploited to assist the user during critical phases or to manage his interaction without interrupting the task at hand or generating mental workload.

However, it should also be noted that subliminal stimulation is affected by a variety of weaknesses described above, such as that of the fast fading of subliminal stimulation's effect. Research in this field is just starting, and many new strategies should still be tested, such as the use of different unconscious stimulation techniques.

The purpose of the present paper was to give an overview of both the most recent findings about subliminal stimuli applied to symbiotic systems and the main stimulation techniques to obtain unconscious perception. Techniques reviewed here are the binocular rivalry, visual masking, visual crowding, and rapid serial visual presentation with some of the latest variants of these classic paradigms.

4.1 Ethics

The use of subliminal stimuli to influence the users' behavior raises some crucial issues related to ethics. Studies reviewed here demonstrated that, under certain conditions, humans' decisions and behaviors can certainly be affected by information not entering the awareness, resulting in a hidden persuasion. Even though the goal of using subliminal persuasion in symbiotic systems is the improvement of the interaction, the question concerning whether the users should be aware that the system uses subliminal stimuli arises.

According to Smidt [43], who argues that the voluntariness of behavioral changes is an essential ethical requirement for persuasive technologies, we believe that users should always be aware of systems using subliminal persuasion. When the user knowingly agrees to interact with a system that uses subliminal stimuli to assist him during the interaction, the assistive, covert influence that eventually could be exerted is not in contrast with the user's voluntariness previously expressed.

Acknowledgements. This research was supported by the European Project CEEDS, The Collective Experience of Emphatic Data Systems, European Integrated Project (Number: 258749; Call: ICT 2009.8.4 Human-Computer Confluence).

References

1. Kouider, S., Dehaene, S.: Levels of processing during non-conscious perception: a critical review of visual masking. Philoso. Trans. Roy. Soc. B Biolo. Sci. **362**(1481), 857–875 (2007)
2. Debner, J.A., Jacoby, L.L.: Unconscious perception: attention, awareness, and control. J. Exp. Psychol. Learn. Mem. Cogn. **20**(2), 304 (1994)
3. Pizzi, D., Kosunen, I., Viganó, C., Polli, A.M., Ahmed, I., Zanella, D., Cavazza, M., Kouider, S., Freeman, J., Gamberini, L., Jacucci, G.: Incorporating subliminal perception in synthetic environments. In: Proceedings of the 2012 ACM Conference on Ubiquitous Computing (UbiComp 2012), pp. 1139–1144, (2012)
4. Cetnarski, R., Betella, A., Prins, H., Kouider, S., Verschure, P.F.: Subliminal response priming in mixed reality: the ecological validity of a classic paradigm of perception. Presence Teleoperators Virtual Environ. **23**(1), 1–17 (2014)
5. Aranyi, G., Kouider, S., Lindsay, A., Prins, H., Ahmed, I., Jacucci, G., Negri, P., Gamberini, L., Pizzi, D., Cavazza, M.: Subliminal cueing of selection behavior in a virtual environment. Presence Teleoperators Virtual Environ. **23**(1), 33–50 (2014)
6. Jacucci, G., Spagnolli, A., Freeman, J., Gamberini, L.: Symbiotic interaction: a critical definition and comparison to other human-computer paradigms. In: Proceedings of Symbiotic 2014: Third International Workshop on Symbiotic Interaction. Springer, Berlin, Heidelberg (2014)
7. Spagnolli, A., Guardigli, E., Orso, V., Varotto, A., Gamberini, L.: Measuring user acceptance of wearable symbiotic devices: validation study across application scenarios. In: Proceedings of Symbiotic 2014: Third International Workshop on Symbiotic Interaction. Springer, Berlin, Heidelberg (2014)
8. Riener, A., Kempter, G., Saari, T., Revett, K.: Subliminal communication in human-computer interaction. Adv. Human-Comput. Interact. **2011**, 58 (2011)

9. DeVaul, R.W., Pentland, A., Corey, V.R.: The memory glasses: subliminal vs. overt memory support with imperfect information. In: Proceedings of the Seventh IEEE International Symposium on Wearable Computers (ISWC 2003), pp. 146–153 (2003)
10. Merikle, P.M., Smilek, D., Eastwood, J.D.: Perception without awareness: perspectives from cognitive psychology. Cognition **79**(1), 115–134 (2001)
11. Tzelgov, J.: Specifying the relations between automaticity and consciousness: A theoretical note. Conscious. Cogn. **6**(2), 441–451 (1997)
12. Dehaene, S., Naccache, L.: Towards a cognitive neuroscience of consciousness: basic evidence and a workspace framework. Cognition **79**(1), 1–37 (2001)
13. Chalfoun, P., Frasson, C.: Subliminal cues while teaching: HCI technique for enhanced learning. Adv. Human-Comput. Interact. **2011**, 2 (2011)
14. McNamara, A., Bailey, R., Grimm, C.: Improving search task performance using subtle gaze direction. In: Proceedings of the 5th Symposium on Applied Perception in Graphics and Visualization, pp. 51–56 (2008)
15. Bailey, R., McNamara, A., Sudarsanam, N., Grimm, C.: Subtle gaze direction. ACM Trans. Graph. **28**(4), 100 (2009)
16. Barral, O., Aranyi, G., Kouider, S., Lindsay, A., Prins, H., Ahmed, I., Jacucci, G., Negri, P., Gamberini, L., Pizzi, D., Cavazza, M.: Covert persuasive technologies: Bringing subliminal cues to human-computer interaction. In: Persuasive Technology, pp. 1–12 (2014)
17. Faivre, N., Berthet, V., Kouider, S.: Nonconscious influences from emotional faces: a comparison of visual crowding, masking, and continuous flash suppression. Front. Psychol. **3**, 129 (2009)
18. Necker, L.A.: Observations on some remarkable optical phaenomena seen in Switzerland; and on an optical phenomenon which occurs on viewing a figure of a crystal or geometrical solid. Philos. Mag. J. Sci. **1**, 329–337 (1832)
19. Breese, B.B.: On inhibition. Psychol. Monogr. **3**, 1–65 (1899)
20. Leopold, D., Logothetis, N.: Multistable phenomena: changing views in perception. Trends Cogn. Sci. **3**, 254–264 (1999)
21. Blake, R.A.: primer on binocular rivalry, including current controversies. Brain and Mind **2**(1), 5–38 (2001)
22. Fox, R., Herrmann, J.: Stochastic properties of binocular rivalry alternations. Percept. Psychophys. **2**, 432–436 (1967)
23. Wolfe, J.M.: Reversing ocular dominance and suppression in a single flash. Vis. Res. **24**, 471–478 (1984)
24. Wilke, M., Logothetis, N.K., Leopold, D.A.: Generalized flash suppression of salient visual targets. Neuron **39**, 1043–1052 (2003)
25. Bonneh, Y.S., Cooperman, A., Sagi, D.: Motion-induced blindness in normal observers. Nature **41**, 798–801 (2001)
26. Tsuchiya, N., Koch, C.: Continuous flash suppression reduces negative afterimages. Nat. Neurosci. **8**, 1096–10101 (2005)
27. Breitmeyer, B.G., Ganz, L.: Implications of sustained and transient channels for theories of visual pattern masking, saccadic suppression, and information processing. Psychol. Rev. **83**, 1–36 (1976)
28. Crawford, B.H.: Visual adaptation in relation to brief conditioning stimuli. In: Proceedings of the Royal Society of London. Series B, Biological Sciences 134, 283–302 (1947)
29. Macknik, S.L., Livingstone, M.S.: Neuronal correlates of visibility and invisibility in the primate visual system. Nat. Neurosci. **1**, 144–149 (1998)
30. Korte, W.: Über die gestaltauffassung im indirekten sehen. Zeitschrift für Psychol. **93**, 17–82 (1923)

31. Levi, D.M.: Crowding–an essential bottleneck for object recognition: a mini-review. Vis. Res. **48**, 635–654 (2008)
32. Bouma, H.: Interaction effects in parafoveal letter recognition. Nature **226**, 177–178 (1970)
33. Martelli, M., Science, N., Majaj, N.J., Pelli, D.G.: Are faces processed like words? A diagnostic test for recognition by parts. J. Vis. **5**(1), 6 (2005)
34. Pelli, D.G., Tillman, K.A.: The uncrowded window of object recognition. Nat. Neurosci. **11**, 1129–1135 (2008)
35. Whitney, D., Levi, D.M.: Visual crowding: a fundamental limit on conscious perception and object recognition. Trends Cogn. Sci. **15**, 160–168 (2011)
36. Kouider, S., Berthet, V., Faivre, N.: Preference is biased by crowded facial expressions. Psychol. Sci. **22**(2), 184–189 (2011)
37. Raymond, J.E., Shapiro, K.L., Arnell, K.M.: Temporary suppression of visual processing in an RSVP task: an attentional blink? J. Exp. Psychol. Hum. Percept. Perform. **18**, 849–860 (1992)
38. Trick, L.M., Pylyshyn, Z.W.: What enumeration studies can show us about spatial attention: evidence for limited capacity preattentive processing. J. Exp. Psychol. Hum. Percept. Perform. **19**, 331–351 (1993)
39. Xu, Y., Chun, M.M.: Dissociable neural mechanisms supporting visual short-term memory for objects. Nature **440**, 91–95 (2006)
40. Shapiro, K., Driver, J., Ward, R., Sorensen, R.B.: Priming from the attentional blink: A failure to extract visual tokens but not visual types. Psychol. Sci. **8**, 95–100 (1997)
41. Luck, S.J., Vogel, E.K., Shapiro, K.L.: Word meanings can be accessed but not reported during the attentional blink. Nature **383**, 616–618 (1996)
42. De Bruijn, O., Spence, R.: Rapid serial visual presentation: a space-time trade-off in information presentation. In: Proceedings of the Working Conference o Advanced visual interfaces, pp. 189–192 (2000)
43. Smids, J.: The voluntariness of persuasive technology. In: Bang, M., Ragnemalm, E.L. (eds.) Persuasive 2012. LNCS, vol. 7284, pp. 123–132. Springer, Heidelberg (2012)

Example Applications

OUTMedia – Symbiotic Service for Music Discovery in Urban Augmented Reality

Pirkka Åman[1(✉)], Lassi A. Liikkanen[2], Giulio Jacucci[2],
and Atte Hinkka[2]

[1] Media Lab Helsinki, Department of Media, School of Art,
Design and Architecture, Aalto University, P.O. Box 31000, 00076 AALTO
Espoo, Helsinki, Finland
pirkka.aman@aalto.fi
[2] Helsinki Institute for Information Technology HIIT,
Aalto University, P.O. Box 31000, 00076 AALTO Espoo, Helsinki, Finland
{lassi.liikkanen,giulio.jacucci,atte.hinkka}@hiit.fi

Abstract. The rise of the mobile Internet has led to emergence of location-based services, but not to commercial breakthroughs in media applications. We created OUTMedia, a location-sensitive music discovery application to investigate the desirable features of user interface in this context. This paper documents our design efforts and a field study using a functional prototype. We utilized several measures in an experiment involving eighteen music enthusiasts. Our findings call for service designers to support the symbiotic interplay between media and places for enriching urban cultural experiences with user-created content. These design implications can support serendipitous media experiences in content discovery services to come.

Keywords: Music discovery · Location-based services · Urban computing · Serendipity · Music interaction

1 Introduction

Discovering meaningful media can be a daunting task for the 21st-century consumer. Internet-based distribution has opened a radically transformed field for consuming media – for both good and bad. Media is available in excess quantities, but more choice will not increase satisfaction if it is too difficult to find the desired content among millions of options. As a remedy, filtering and recommendations have previously been used mainly in desktop applications and localization information in mobile applications, but this is currently undergoing a change, and the future promises ubiquitous context-sensitive media recommendations and discovery technologies [1, 5, 6, 17–19].

Location-based, context-sensitive services seem particularly suitable for recommending recorded and live music. Live music has a natural connection to the physical environment where it is performed. Listening to recorded music in a public setting with headphones allows a person to create a private auditory bubble [7]. Beyond new experiences and empowerment in urban space realised by the potential of the ubiquitous digital media [24], location-sensitive music could be a source of serendipity [8].

© Springer International Publishing Switzerland 2014
G. Jacucci et al. (Eds.): Symbiotic 2014, LNCS 8820, pp. 61–71, 2014.
DOI: 10.1007/978-3-319-13500-7_5

As the relation of music and location can be arbitrary, this allows great opportunities for random encounters of good or useful content unexpectedly, i.e., serendipity. This has been the goal of many music recommendation systems [27] and even in music listening [15], but without considerations of the mediating role of physical location for the experience.

We have previously studied the feasibility of urban music and event recommendations in a map-based design [10]. We wanted to expand this approach to look into richer associations of music and place with a taste of social computing. Because there were no popular location-based music services available (with enough users or content) we were required to simulate the (user-created) content. We decided to do this in a custom-built research application using an augmented reality (AR) interface that links music content with a specific place. AR allows for a direct connection and physicality with the space, unlike using a traditional 2D map interface. AR solutions are also natively *location-sensitive*, i.e. they commonly deploy a user's physical location for interaction. As further motivation, our review shows that current context-aware music apps do not use AR, and urban AR apps do not support music discovery. Our application connects music to places freely chosen and annotated by users.

In this paper, we report an exploration into location-sensitive media discovery. We developed a functional prototype of a location-based music service, OUTMedia, through a research-through-design approach [9]. We deployed OUTMedia with simulated user data and music objects (in a Wizard-of-Oz fashion) to study user experiences (UX) and understand experiential design success factors. We were interested to learn how active music listeners experienced this service in terms of serendipity, overall UX, and engagement. This combination of research goals and novel technology makes our contribution unique in the literature. Our results indicate that the main design goals were fulfilled, and our 18 participants maintained sustained attention to the application and enjoyed freely discovering new music, as measured by the adapted ResQue instrument [25]. OUTMedia clearly supported symbiotic experiencing of place, time and media content through ubiquitous human-computer interaction.

2   Background and Related Work

The previous work that is most relevant to our study can be found in two fields: contextual music recommendations and location based services. There are many commercial and research mobile AR applications, but since development of new mobile AR interactions is not the main focus of our study, we will not review them in depth.

2.1   Music Discovery and Urban-Augmented Reality

Traditional music recommendation services take two forms: music or other users and their musical preferences [1] can be recommended. This ignores the fact that music choices depend heavily on a user's situation [16, 18]. With the popularity of mobile devices with rich sensing capabilities, researchers have started to employ contextual information in mobile music recommenders [6, 26]. Context-aware music recommender applications employ a third dimension: different contextual factors [1], such as location, time, function (e.g. sports, reading), and mood, which may influence the user's preferences at a particular moment [14, 19]. In a recent paper, [2] discuss several location-based music service concepts, which are still mostly unrealized.

AR applications add new digital elements into the physical world using a computing device. AR applications are expected to gradually become part of everyday computing [21]. With AR, local searching no longer means looking down at a screen, but also looking out at the world in a more natural way [20]. Currently, AR media is approaching mainstream computing through products such as Google Glass and Nokia City Lens.

According to our review, OUTMedia is the only one that (a) utilizes context sensitivity (through an AR interface), (b) offers music as its main content type, and (c) lets users create the located content freely with their own annotations. Of the most closely related research [3, 4, 6, 10, 13] and commercial (Tunaspot [www.tunaspot.com], Spotisquare [www.spotisquare.com], Soundtracking [www.soundtracking.com], Layar [www.layar.com], Wikitude [www.wikitude.com]) applications, Tunaspot resembles OUTMedia, but it is not context-sensitive in the manner we use the concept: that *content can only be consumed in close proximity to the location.*

Fig. 1. Augmented reality view of the OUTMedia browser.

3   OUTMedia App: Design and Prototype

We had studied urban music discovery with a map-based approach earlier and now wanted to study the user perception of AR UI. We used UI solutions that have proven already successful in commercial applications such as Layar and Wikitude. These apps have tested and proven design elements and features.

The concept of OUTMedia application was developed through a series of user workshops. The concept of a location-sensitive music service included not only the idea that music objects have location, but users can also create objects and provide their own annotations for music-place combinations for others to discover. We were also interested in how curated, location-based playlists functioned *in situ*. In our application, user first scans her surroundings for finding interesting 'floating' AR objects, then taps the object to open it, revealing a media player that contains also a field for varying textual annotations.

The novel feature of our system is in offering location-sensitive music content that is attached to places according to various reasons. We included user-created, promotional and automatically generated content and user-created annotations *together* with presenting them through an AR UI. We tested our concept by building a prototype that realized some features of the concept.

3.1   AR Objects and POIs

The AR objects were the primary focus of interaction in OUTMedia. They were associated with geographical POIs. Each AR object had seven properties: *POI, category, music clip, time stamp, title, text annotation*, and a *user name tag*.

Most of the POI locations were gathered from users in the workshops. Users were asked to mark on the map of the city district of the study (a) where they spend time regularly, (b) where they would leave music and (c) why would they do so. However, the final locations, all textual information, and the music were selected and created by the researchers, resulting in an adapted Wizard-of-Oz study [12].

We had four categories of AR objects: user-created content (UCC), music events (live gigs and DJ events), and promotional (food or drink). These were marked with different colours and icons. In addition to categorized objects, OUT Media supported 'Soundtrack of a place' feature. It simulated automatically generated playlists of the four latest plays around the place.

The text annotations in the different categories were semantically distinct and always under 140 characters. For music events and promotions, annotations were knowledge- based, describing the performer or detailing the promotion. In the soundtrack, the annotation contained the playlist.

User-created annotations represented two types: *knowledge- based* and *arbitrary* (cf. tag-based relation in [6]). In knowledge-based objects, there was a factual relationship between a place, music, and annotation. For instance, '*Find punks here*' annotation was made for a local punk song at a POI usually occupied by punks. Arbitrary relation simulated user activity by linking music, a place, and annotation by what feels right or wrong (ironic, controversial, etc.).

3.2   The User Interface

There were two views in the user interface: an AR view (Fig. 1) and a player view. In the AR view, users saw objects floating around the screen. After the user touched an object in the AR view, the player view opened. The player view presented the user an audio player, which included a timeline, a play/pause button, and Favourite button. The time stamp, annotation, and user name were also shown.

In addition, the AR view had two controls on the upper right corner, 'Friend filter' and 'Soundtrack of the place,' and a play and pause button on the lower right corner. The Friend filter function mimicked social cues and filtered visible objects based on the real interactions of the user's chosen 'Friend.' Activating the filter limited the count of visible objects to eight. Chosen first on the basis of the items the friend liked, then the items the friend played, and last the items the friend opened in the player.

3.3 Generating Content for the Field Study

We generated over one hundred ad hoc POIs and AR objects to mimic user created content. Of the AR objects, 60 % were user-created (UCC) and 40 % belonged to the other categories to simulate a realistic divide between various categories. The POIs were distributed evenly across four Helsinki parks, each having one focal spot around which they were spread. We chose the district known for its density of bars and outdoor hangouts. There were on average 28 objects in each cluster, totalling 111 objects.

Researchers knew the chosen district thoroughly and selected the music and POIs to match the taste of target segment users that spent time or lived in the area. The application was otherwise mostly functional, but instead of free movement, it supported only four pre-defined clusters where the selected objects were visible.

The management of the POIs and AR objects was done in a web application that provided batch output for an Android application assembly phase. The motivation to use the pre-assembled collection of data embedded in the application was the need to guarantee a consistent UX.

3.4 Implementation

The application was developed for an Android platform and the experiments were conducted on Samsung S3 devices. The device has a 4.8-inch screen and includes an 8 MP camera that provides the image for the AR application. Determining AR object visibility is a three-stage decision process. (1) Does the object belong within the field of view of the camera? (2) Is the object part of the cluster the user is in, i.e., park 1–4 (determined by the distance to the cluster's location)? (3) Is the Friend filter on or off?

The application relies on the smartphone accelerometer and compass to determine the orientation of the camera of the device. It determines the physical field of view for the camera of the device and uses this information to calculate the visible objects. Only the compass angle is taken into account when determining visibility and the objects are placed on an imaginary Cartesian plane. The placement algorithm in the application is straightforward. On the X-axis, the location of the object is determined from its deviance from the camera-pointed angle; on the Y-axis, by its distance from the user compared with the other visible objects. The nearest object is drawn at the bottom of the screen, and the other objects behind it and on the Y-axis on top of it. To prevent the drawn items from moving too quickly for the user and to provide an illusion of them moving, the location of a given item is determined by linearly interpolating its location, given its location in the previous frame and its ideal location calculated by the placement algorithm.

4   Method

We organized a field experiment to evaluate our location-sensitive music discovery service. There are several advantages of doing field research with mobile applications [11, 23]. However, real-life settings also pose disadvantages including traffic, the weather, outsiders, and other noise sources. We considered a field study a necessity, also because of limited application functionality that did not allow for handing it for private use.

The focus of the study was to evaluate the acceptance of the prototype. This included the persuasiveness of different kinds of AR objects providing music recommendations. Our main research questions were as follows: Does the system *engage users in music interaction*? Does it provide *serendipitous music recommendations*? Does it *inspire users' experience of the places through music content*? To answer the questions, we performed interviews, used structured surveys and gathered log data.

4.1   Participants and Experiment Design

We recruited 18 participants using snowballing through social media. The following criteria were used: 'active seekers of new music, active listeners of music using mobile devices, and familiar with smart phones. They were 23–41 years of age, 10 female and 8 male. In the recruitment phase, the first 9 participants were asked to name a friend of similar criteria for the testing of the social features of the application (Friend filter). Altogether, we had 9 pairs of participants. All were familiar with the district of the experiment and its publicly known sites and locations.

All users were free to use the app in the four sites. One quasi-experimental manipulation was also administered. In two of the four parks, half of the users were urged to use the *Friend filter*, which allowed them to see only those eight AR objects that their friend-pair had liked, played, or opened. If the friend had liked, played or opened more than eight objects, they were filtered in the order of interaction depth: like over play, play over open.

4.2   Experiment Procedure

The study started with a pre-experiment briefing, where users were informed that they could listen to the tracks as long as they want, but it was advisable to spend around 15 min in one park. They were also told that the route of four parks would take about one and a half hours and right after that a questionnaire and an interview would be administered. Upon inquiry, users were told that the POIs were gathered from users like themselves using a printed map. Users were informed that the radius for visible POIs was around 250 m, so they were in eyesight or within a short walking distance.

Users were able to 'Favourite' the POIs they found interesting. After presenting orientation material, users were given the Android smartphone with the application and Bluetooth headphones.

We provided 111 AR objects in four geographical clusters and within walking distance (400 m) from each other. These four clusters formed a route approximately 2 km long. The experiment started at the southernmost park with the guidance of the researchers and proceeded to the northernmost, where the interviews took place and the survey was completed. Upon arrival, participants were free to use the application to scan their surroundings for AR objects, listen to the songs, and interact with the app. All participants were able to interact with all 111 objects. Participants were asked if they would be ready to move on to the next park after they had spent 20 min in one park and had not themselves indicated the desire to move on.

4.3   User Evaluation

We used a combination of methods for evaluating user experience. These included analyses of log data, a questionnaire, and interviews. In this paper, we focus on behavioural and usability-related topics.

The OUTMedia application logged all user interactions and system statuses in the absence of interaction. All user interactions were recorded sequentially with a new timestamp applied every second. The log data was pre-processed using Microsoft Excel and analysed with SPSS.

After the experiment, users completed a questionnaire, which had 17 claims about (1) *User-perceived qualities,* six claims about (2) *User beliefs,* one claim about (3) *User attitudes,* and four claims about (4) *Behavioural intentions.* Claims were operationalized from constructs such as User attitudes (e.g. *'Overall, I am satisfied with the recommended items'*) or Behavioural intentions (e.g. *'If a recommender such as this exists, I will use it to find products to buy'*). The ResQue instrument, standing for *Recommender systems' Quality of user experience,* is designed to measure the whole user experience of recommender systems, including user-perceived qualities and satisfaction levels [25].

Participants were interviewed after the experiment about 12 themes that explored the concept's and application's acceptability and UX, the perception of various media types within the objects, and quality and contents of POIs.

5   Findings

This section describes the behavioural and usability results of our study. We will return to design implications in the Discussion.

When asked if they would like to scan their surroundings this way for discovering other things, 67 % of the users felt that music together with user comments or curated information was enough for the experience. Some users mentioned that photos or videos might be too sense-consuming in an outdoor setting, while others felt that it could result in an improved experience.

5.1 Behaviour and Usability

We analysed data logs from all 18 subjects. On average, user sessions took 1 h 25 min, with considerable variation between subjects (S.D. = 37 min). Users utilized the music browsing extensively. On average, each user browsed through 56 POI items (S.D. = 22 items; opening the player window), almost exactly 50 % of all available POIs. They started listening to 35 tracks (S.D. = 12.3 tracks). This indicates that over two-thirds (67.1 %) of opened POIs were also listened to. Users also interacted with the player, marking on average 19 tracks (S.D. = 9 tracks) with a star and removing one 'Favourite' per session (M = .83; S.D. = .86).

Attractiveness of POIs. The most played and liked POI category was 'Soundtrack of a place.' The soundtracks differed from other categories in that they were not moving objects like other POIs, and had their own symbol in the upper right corner of the UI (Fig. 1). Users also played back Soundtracks for longer than other tracks (1:12 vs. 0:51, $F(1, 633) = 5.622$, $p = .018$). This is related to the fact that soundtracks consisted of multiple tracks and the user had the additional control element 'Skip' available. This button was used on average four times per session although not all users chose to wind forwards (M = 4.1; S.D. = 4.8). There were no significant differences between other types of POI categories.

Of the other categories, survey data shows that users liked UCC, Live gigs and DJ objects most, followed by soundtracks and promotional objects. Live gigs and DJ categories had timely information about what was happening within walking distance. That explains their high rank among categories. Survey data shows also that user-created POIs were the most liked. However, log data shows that UCC POIs were played and liked as much as all categories on average. Interestingly, the interviews revealed that in the case of UCC, users felt that even when the music was bad, if the comment was interesting they chose to like the POI, and vice versa. Log data shows that in 10 % of the most played (N = 12) and most liked POIs (N = 13) there were no differences between categories. Also, over half (58 %) of the most played were present in the most liked group as well.

Influence of Social Information. Enabling the Friend filter influenced interactions with the application only slightly, mostly in terms of the amount of attention given to POIs. On average, POIs were viewed for 14.5 s with the filter, but for 20.0 s without it. The difference was statistically significant despite the remarkable variability in behaviour (T-test not assuming equal variances, $t(135.304) = 2.758$, $p = .007$). This left less time for users to perform other activities, such as liking the POI.

Usability and UX Assessment. The most common feedback of usability in the interviews and during the sessions concerned the AR objects that were located the furthest away. Because they were the smallest and often overlapping (occluding) with other objects, half of the users felt that it was sometimes hard to hit the right object. In the design phase, we decided to scale them so that even the smallest ones could be hit easily, but it seems that the scaling and the movement of the objects could be further optimized. However, usually the complaints happened right after the start of the session and users quickly got the idea of the application. Another common usability-related comment that arose from the interviews was the need for a generic Back button in the AR view. This was felt to be needed for going back to the player view for liking the currently listened song (when already back in the AR view and looking for new objects). According to the ResQue UX metrics, grand average of the perceived user experience resulted just over four on a 1–5 Likert scale.

Desired features and symbol recognition evaluation. Half (9) of the participants expressed wishes that OUTMedia would include more *social media* features. They specifically wished to leave their own comments and see them presented on a Facebook-style timeline. It was also mentioned that meaningfulness of comments depends on social proximity, especially when commenting on or reading close friends' comment.

Finally, we also ensured that the users had acknowledged the category allocations we thought important. In the questionnaire, users were asked to recognize the six symbols of different object categories. Symbol recognition produced 94 % (N = 18) right answers. We conclude that UI symbols were easy to remember and conveyed the intended meanings.

6 Discussion

In this paper, we presented an evaluation of a location-based music discovery service, OUTMedia. In the user study, we found that it fulfilled our main design goals: supported serendipitous music discovery and engaged users in music interaction [17]. We found out that the application use resulted in experiencing time, places, and various media content in new ways.

The application provided several types of serendipity. The popularity of features or content was widely distributed between subjects, showing *exploration* as an aspect of serendipity, as not everyone found or liked similar items. There was also diversity as the questionnaire and interview data revealed multiple reasons for participants using the Favourite feature on POIs. In addition to 'real' serendipity, there was also pseudo-serendipity (re-discovery). Based on interviews and ResQue metrics, the application supported music interaction and discovery very well.

The behavioural data reveals other aspects of interaction. The short average playback times implies that users mostly used the application for browsing and checking out music (recommendations). This was prominent when the Friend filter was activated, which indicates that the presence of social information induced 'status checking' behaviour [22]. That is, the user was not necessarily interested in the POI, but wanted to explore it because their friend had expressed some interest in it. After browsing, they continued checking out the AR scanning mode, instead of using the application in the same manner as when the Friend filter was off. However, Friend filter was perceived positively among users in the interview and survey data. So was the 'Soundtrack of a place', a mix of several songs for the place. This suggests that a 'music trail' type of continuous, radio-like concept would be a topic to explore in future.

One limitation concerns the social media features of the application. Some of the users felt that they would have liked the possibility to leave music in places themselves, and this is precisely what the complete service should enable. However, in the context of a design study, this was unfeasible. Also, during a limited time span, the accumulation of the amount of shared POIs would have been too low to mimic realistic social media services.

7 Conclusion and Future Work

In this study, we discovered some central UX elements for location-based music discovery service. We also presented design insights based on user data for how to design location-based mobile AR services. We stress that the roles of location, time and various media types should all be considered carefully in designing for serendipitous location-sensitive experiences. An interesting future work direction would be to explore the 'extensions of man, restrictions of man' approach with an auditory AR for music and possibly other audio discovery. For example, a simple interface that would allow for discovering and hearing only music or other audio that is left within hearing distance. Sounds and music from the past would come closer and fade out by walking through city streets, resulting in a city layered with aural histories. Regardless of future development, we hope the present study and our design suggestion inspires the designers to create better technology mediated experiences for music lovers.

References

1. Adomavicius, G., Tuzhilin, A.: Towards the next generation of recommender systems: a survey of the state-of-the-art and possible extensions. IEEE Trans. Knowl. Data Eng. **17** (6), 734–749 (2005)
2. Åman, P., Liikkanen: L.: Painting the city with music: context-aware mobile services for urban environment. Continuum J. Media Cult. Stud. **27**(4), 542–557 (2013)
3. Ankolekar, A., Sandholm, T., Yu, L.: Play it by ear: a case for serendipitous discovery of places with Musicons. In CHI '13, Paris, 27 Apr to 2 May 2013
4. Balduini, M., Celino, I., Dell'Aglio, D., Valle, E.D., Huang, Y., Lee, T., Kim, S., Tresp, V.: BOTTARI: An augmented reality mobile application to deliver personalized and location-based recommendations by continuous analysis of social media streams. Web Semant. Sci. Serv. Agents World Wide Web **16**, 33–41 (2012)
5. Baumann, S., Jung, B., Bassoli, A. Wisniowski, M.: BlueTunA: Let your neighbour know what music you like. In: CHI '07, San Jose, 28 Apr to 3 May 2007
6. Braunhofer, M., Kaminskas, M., Ricci, F.: Recommending music for places of interest in a mobile travel guide. In: RecSys' 11, Chicago, 23–27 Oct 2011
7. Bull, M.: Sound Moves: iPod Culture and Urban Experience. Routledge, New York (2008)
8. Celma, O.: Music Recommendation and Discovery. Springer, Heidelberg (2010)
9. Fällman, D.: The interaction design research triangle of design practice, design studies, and design exploration. Des. Issues **24**(3), 4–18 (2008)
10. Forsblom, A., Nurmi, P, Åman, P., Liikkanen, L.: Out of the bubble—serendipitous event recommendations at an urban culture festival. In: IUI '12, Lisbon, 14–17 Feb 2012

11. Goodman, J., Brewster, S.A., Gray, P.D.: Using field experiments to evaluate mobile guides. In: Mobile HCI '04, Glasgow, 13–16 Sept 2004
12. Klemmer, S.R., Sinha, A.K., Chen, J., Landay, J.A., Aboobaker, N., Wang, A.: Suede: a wizard of oz prototyping tool for speech user interfaces. In: UIST '00, San Diego, 5–8 Nov 2000
13. Lehtiniemi, A., Ojala, J.: MyTerritory: evaluation of outdoor gaming prototype for music discovery. In: ACM MUM '12, Ulm, 4–6 Dec 2012
14. Lehtiniemi, A.: Evaluating SuperMusic: streaming context-aware mobile music service. In: ACE '08, Yokohama, 3–5 Dec 2008
15. Leong, T.W., Vetere, F., Howard, S.: Experiencing coincidence during digital music listening. ACM Trans. Comput.-Human Interact. 19(1), 1–19 (2012)
16. Levitin, D.: Life soundtracks: the uses of music in everyday life. McGill University, Montreal (2007)
17. Liikkanen, L.: Music interaction research in HCI—Let's Get the Band Back Together! Paper presented in CHI '12, Austin, 5–10 May 2012
18. North, A.C., Hargreaves, D.J., Hargreaves, J.J.: Uses of music in everyday life. Music Percept. Interdisc. J. 22(1), 41–77 (2004)
19. Nurmi, P., Bhattacharya, S.: Identifying meaningful places—the nonparametric way. In: IEEE Pervasive '08, Hong Kong, 17–21 Mar 21 2008
20. Olsson T., Väänänen-Vainio-Mattila, K.: Expected user experience of mobile augmented reality services. In: MobileHCI '11, Stockholm, 30 Aug to 2 Sept 2011
21. Olsson, T., Salo, M.: Online user survey on current mobile augmented reality applications. In: ISMAR '11, Basel, 26–29 Oct (2011)
22. Oulasvirta, A., Rattenbury, L., Mai, L., Raita, E.: Habits make smartphone use more pervasive. Pers. Ubiquit. Comput. 16(1), 105–114 (2012)
23. Oulasvirta, A.: Field experiments in HCI: promises and challenges. In: Saariluoma, P., Roast, C., Punamäki, H.K. (eds.) Future Interaction Design: Part 2. Springer, New York (2008)
24. Paay, J., Kjeldskov, J., Howard, S., Dave, B.: Out on the town: a socio-physical approach to the design of a context-aware urban guide. ACM Trans. Comput-Human Interact. 16(2), 7–34 (2009)
25. Pu, P., Chen, L.: A user-centric evaluation framework for recommender systems. In: RecSys '10, Barcelona, 26–30 Sept 2010
26. Wang, X., Rosenblum, D., Wang, Y.: Context-aware mobile music recommendation for daily activities. In: ACM MM '12, Nara, 29 Oct to 2 Nov 2012
27. Zhang, Y.C., Séaghdha, D.Ó., Quercia, D., Jambor, T.: Auralist: introducing serendipity into music recommendation. In: ACM WSDM '12, Seattle, 8–12 Feb (2012)

Symbiotic Wearable Robotic Exoskeletons: The Concept of the BioMot Project

J.C. Moreno[1]([⊠]), G. Asin[1], J.L. Pons[1], H. Cuypers[2], B. Vanderborght[2],
D. Lefeber[2], E. Ceseracciu[3], M. Reggiani[3], F. Thorsteinsson[4], A. del-Ama[5],
A. Gil-Agudo[5], S. Shimoda[6], E. Iáñez[7], J.M. Azorin[7], and J. Roa[8]

[1] Neural Rehabilitation Group, Cajal Institute,
Spanish Research Council, Madrid, Spain
jc.moreno@csic.es
[2] Department of Mechanical Engineering, Vrije Universiteit Brussel, Brussel, Belgium
[3] Department of Management and Engineering, University of Padua, Padua, Italy
[4] Ossur Hf, Reykjavík, Iceland
[5] National Hospital for Spinal Cord Injury, Toledo, Spain
[6] Intelligent Behavior Control Collaboration Unit, RIKEN, Wako, Japan
[7] Biomedical Neuroengineering Group,
Univ. Miguel Hernández de Elche, Alicante, Spain
[8] Technaid SL, Madrid, Spain
http://biomotproject.eu

Abstract. Wearable robots (WR) are person-oriented devices, usually in the form of exoskeletons. These devices are worn by human operators to enhance or support a daily function, such as walking. Most advanced WRs for human locomotion still fail to provide the real-time adaptability and flexibility presented by humans when confronted with natural perturbations, due to voluntary control or environmental constraints. Current WRs are extra body structures inducing fixed motion patterns on its user. The main objective of the European Project BioMot is to improve existing wearable robotic exoskeletons exploiting dynamic sensory-motor interactions and developing cognitive capabilities that may lead to symbiotic gait behavior in the interaction of a human with a wearable robot. BioMot proposes a cognitive architecture for WRs exploiting neuronal control and learning mechanisms the main goal of which is to enable positive co-adaptation and seamless interaction with humans. In this paper we present the research that is conducted to enable positive co-adaptation and more seamless interaction of humans and WRs.

Keywords: Bionspiration · Neurorehabilitation · Robotics · Biomechanics · Neuromusculoskeletal models · Compliant actuation

This work is partially supported by the Commission of the European Union. FP7-ICT-2013.2.1-611695.

G. Jacucci et al. (Eds.): Symbiotic 2014, LNCS 8820, pp. 72–83, 2014.
DOI: 10.1007/978-3-319-13500-7_6

1   Introduction

The possibility of interfacing the human body with artificial devices to then employ these devices to assist human function (e.g. enhance performance, restore neurological function), has long fascinated mankind. The first examples of such interactive devices were industrial robots that had been adapted and brought into manufacturing environments. In the context of assistive technology, Wearable Robots (WRs) were introduced in [1]. WRs, e.g. exoskeletons, are person-oriented robots that directly interact with the user to supplement the function of a limb. Over the last two decades remarkable technical accomplishments in design methods have been achieved and have led to a few commercialized products. Despite these technological and mechanical advances and their use in the clinical environment, key issues related to the implementation of walking WRs in daily life have largely been ignored. The approaches to man-machine interaction put forward by the most recent walking WRs need to consider effective strategies for interfacing a WR to the human body based on the interplay between the neural and the musculoskeletal system [2].

1.1   Concept

One way to characterise and establish the symbiotic relations between humans and WRs is to synthesize the walking behavior and adaptation. In this characterisation we set a theoretical framework to understand walking by means of a collection of models of diverse variety and hierarchy level (cortical, neuromuscular, mechanical). The coordination of the cooperation within and between the

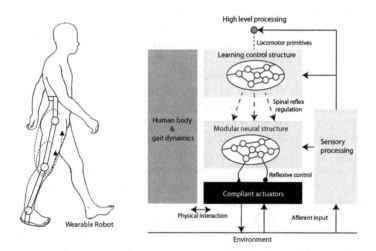

Fig. 1. BioMot concept for symbiotic human-exoskeleton interaction. The different control levels of the artificial system are represented: interplay between biomechanical (compliant structures) and sensory-motor levels (spinal and low level structures) with a developmentally guided coordination (learning structure).

different functional levels of the motor system, including motor learning, could result in successful developmentally guided coordination between neuronal activity and the biomechanics of the musculoskeletal system.

So-called intelligent machines are today far from the flexibility found in the real-time adaptability of humans when confronted with changes in environmental (uneven terrain, mechanical perturbations, carrying weight) and task (transitions, changes in walking speed and direction) constraints. Bridging the adaptability gap requires a further comprehension of the globally organized mechanism for integration of action and perception.

Such global organization for WRs is inspired by the neurophysiological mechanism of animal locomotion, in which the interaction between central pattern generators and the sensory system generates an adaptive locomotion. Neural basis of biped locomotion and the mechanisms that are underlying the adaptation to environmental and task constraints are yet to be exploited in a new generation of adaptive WRs. Design of artificial controllers for walking bipeds and exoskeletons has relied to a major extent on empirical work and validation with physical bench and prototype testing. However, novel real-time neuromusculoskeletal computer-based simulations of the human lower extremity in interaction with a WR are substantial for comprehension of the properties of the neural and musculoskeletal systems during a dynamic interaction. In this context, understanding the mechanisms that mediate the cognitive anticipatory processes and the contributions of cognitive function to changes in task constraints are an important requirement for the design of robust and safe WRs. Exoskeletons for gait rehabilitation have been developed for adaptable compliance in robot-human interaction during fixed and steady conditions. In order to move WRs into unconstrained daily-life environments, adaptable compliance needs to be developed to adjust the body dynamics to the desired motions. This concept has been explored mostly with walking and running robots, in which compliance has shown the possibility to extend the capabilities of these devices. For instance, the EU grant ESBIRRO had as a major goal the biomimetic control of walking to build autonomous walking bipeds; the RunBot Project exploited the structural coupling of neural and mechanical levels for a flexible biped robot. A cognitive architecture exploiting neuronal control and learning mechanisms for WRs is proposed (see Fig. 1). In our concept we aim to adjust the particular dynamics of wearable exoskeletons and to exploit bio-inspired strategies for co-adaptation.

This framework is conceived to investigate and build a truly real time cognitive system of a WR, based on a neurophysiological mechanism for animal locomotion, that maintains a stable locomotion against environmental and task constraints. The main goal of BioMot is to exploit dynamic sensory-motor interactions to develop complex cognitive capabilities that can lead to symbiotic gait behavior, producing improved systems and providing guidelines and benchmarking to improve future systems. In this paper we present the research that we are conducting to achieve robust methods for extracting information from the movement dynamics while navigating and performing daily activities and introduce how we are analyzing the use of this information for direct interaction with wearable assistive robots that can train or/and support neurologically injured subjects (e.g. spinal cord injuries). The research is designed to comprehend dynamic sensory-motor interactions in realistic human locomotion that can be applied for design of artificial cognitive systems embodied into the WRs.

2   Research Workpackages

The BioMot project proposes a framework to fuse the information from both interaction with the environment and human gait dynamics, and exploit this information for seamless interaction and safe locomotion adjusted to the user's intentions and capabilities. Experimental analysis of the gait process are conducted along Workpackage (WP) 1, using existing wearable exoskeleton technologies. These experiments set the basis for the development of algorithms to implement sensory-motor loops and regulation of reflexes at the biomechanical and spinal level in WP3, tested in Neuro-Musculo-Skeletal (NMS) simulators. These control levels are transferred to physical simulators (with compliant structures developed in WP2) for partial validation and demonstration in WP4. Then, these are finally integrated in WP6 with higher-level learning algorithms (developed in WP5) to constitute the smart WRs. The background of the research topics addressed by the project's WPs and our research hypotheses and goals are presented in the following sections.

2.1   Walking with Wearable Exoskeletons

Robotic lower limb exoskeletons hold significant potential for gait assistance and rehabilitation; however, there is still a limited understanding of how people adapt to walking with robotic devices. Preliminary research has shown that user adaptation to robotic assistance reveals a reduction in net muscle moments about the joint, while the total (combined) joint moment remains closer to the ones obtained while walking without robotic assistance, altering joint kinematics [4]. However, it has not been clearly attributed to subject's adaptation or actuator mechanical limitations, such bandwidth and power capacity. Further physiological adaptation has been seen in the whole leg stiffness during hopping, in which the biological leg stiffness is modulated in order to achieve a combined linear stiffness, which is a characteristic of normal hopping, with a reduction in energy consumption. Besides, the choice of orthosis control method can greatly alter how humans adapt to robotic assistance during walking. Experiments done with an ankle-foot robot during walking have shown that bio-inspired, proportional myoelectric control, results in larger reductions in muscle activation and gait kinematics more similar to normal compared to mechanistic footswitch control [5]. However, co-contractions of main ankle muscle provide co-contraction of robot actuators also, hindering gait kinematics. Regarding whole leg biomechanics during walking, there is little evidence of exoskeletons reducing the metabolic cost of walking [2] or the effect of an exoskeleton on the agility of the user's movements to adapt to changes such as walking speed, turning, etc. Preliminary research on those topics has shown that manipulation of leg impedance through the exoskeleton increases leg bandwidth, which would enable the user to impart greater accelerations to the leg and, improving the operators's capability to perform corrective movements to maintain balance or to effectively reduce the metabolic cost during walking [6]. Robotic functional compensation during walking is a challenging field that requires design of compliant, highly back-drivable actuators and control strategies that provide joint assistance and stabilization and unhindered movement to maximize user residual capabilities. Despite the technological advances made in military and transferred to clinical applications of walking exoskeletons, there is still a frontier that has not been successfully overcome yet, which is the transition from stationary robotic exoskeletons to ambulatory robotic exoskeletons. Comfortable use with commercially available exoskeletons, such as the Ekso or Rewalk, is still distant from independent use. Large periods of expert supervision and tuning for an autonomous use are still required [8]. In this WP firstly, we are producing real-time regulation of joint motion and impedance. We propose neuronal control strategies to realize agonist-antagonist control structure of lower limb joints with controllable exoskel etons. The proposed local sensor neurons can achieve inter-joint control, measuring peripheral information and influencing joint movements with actuation mechanisms. Secondly, we investigate the discrimination between locomotion modes. A trade-off between extrinsic and intrinsic control modalities is assumed to design a sensor-based control of the robotic joint systems. The interface relies on EMG and internal kinetic sensors for intrinsic control of exoskeletons with seamless interaction.

2.2 Motor Programmes for Control of Bipedal Walking

There is evidence indicating the existence of a patterned control of human locomotion [9]. This means that instead of controlling single muscles and degrees of freedom individually, coordination of movements for gait is achieved by small repertoire of rhythmic signals. These are so called activation patterns (APs), which are delivered to several different muscles involved in a single motor task based on feed-forward and feedback signals, and as a function of the phase. These APs are produced by neural networks, located primarily in the spinal cord and are also known as central pattern generators (CPGs). Thus, the activity of each muscle involved in the motor task is the result from a weighted combination of all APs. Several studies have shown that the EMG activity of trunk and leg muscles during human adult locomotion is adequately reconstructed as a linear combination of four to five basic APs, each one timed at a different phase of the gait cycle [10]. Systematic correlations between the timing of the APs and the occurrence of specific biomechanical events of the gait cycle have been shown. Correlational analyses [9] and biomechanical simulations based on the experimentally derived APs showed the contributions of different basic patterns to the walking sub-tasks (body support, forward propulsion and leg swing); also, evidence of a context-dependent correlation of an additional pattern associated with ilio-psoas and erector spinae muscles, involved in accelerating the leg forward and stabilizing the trunk in late stance and early swing, has been found [13]. Several studies have revealed some adaptation mechanisms of APs to environment changes, for instance, once the average waveform of each AP is time-normalized to the step duration, it is little affected by changes in walking speed, direction, loading or unloading of the limb and body or changes in locomotion mode, however, the timing and the weight of the patterns may change considerably as a function of the mentioned factors. Some authors have studied the effect of sensory feedback on control of different aspects of locomotion, as speed and position, revealing the influence of various sensory receptors [14]. In BioMot we work with computational and physical simulations to test how proprioceptive signals can be integrated into central pattern generators that resemble the basic functional characteristics of these central networks. Biomechanical and neurophysiological measurements on human subjects are taken for a better insight into the biomechanical correlations of APs of bipedal walking, and their mechanisms of adaptation to environment. The experimental scenarios are conceived to confront operators with changes in environmental and task constraints towards a better knowledge of the inherent mechanisms that regulate sensory feedback on the modulation of APs.

2.3   Cognitive and Behavioural Contributions to Gait Dynamics

Simple physical tasks, such as turning a page to access information or walking from a place to another independently, require an intense brain activity. This kind of motor movements occur in milliseconds and require little conscious thinking. High-level commands are generated by the brain and translated into the low-level muscular and nervous actions that are needed to perform a particular movement. One of the most significant current discussions in gait function restoration is how the cognitive function affects the lower limb locomotion during walking. It has been shown that the primary motor cortex carries information about bipedal locomotion [15]. Although walking is automatically based on reflexes governed at the spinal level, there are evidences that suggest that the motor cortex is particularly active during specific phases of the gait cycle [16]. These findings support the idea of finding a relationship between electroencephalographic (EEG) signals and different parameters of the gait cycle. Particularly, Petersen and colleagues found a synchrony in the frequency domain (coherence) between the primary motor cortex and the tibialis anterior (TA) muscle indicating a cortical involvement in human gait function [16]. Also, strong Event Related Desyncronization (ERD) components have been found while performing normal walking [18]. Other studies claim that EEG signals are directly related to the value of joint angles involved in human gait [19]. To that end, hip, ankle and knee angles are decoded using linear regression models. Moreover, the level of attention during walking may infer differences in the corticospinal drive while performing dual-tasks so it could be expected to also find differences in cortical activity during EEG measurements. In BioMot, we aim at understanding motor primitives through the analysis of neurophysiological, biomechanical and perceptual responses using WRs during human locomotion. Experimental studies with healthy and neurologically injured subjects are conducted to determine temporal, frequency and spatial characteristics of the cortical control signals that may show coherence with gait cycle (and transitions) parameters including approaches based on muscle activation analysis. Moreover, BioMot analyses the influence of cognitive tasks in more specific gait functions such as automatic gait initiation, changes in direction/orientation during walking or generation of locomotor patterns in dynamic environments.

2.4   Compliant Actuation of Wearable Robots for Gait

Bipedal robots are typically actuated with stiff servomotors. An encoder is attached to measure the position and a feedback loop (often high gain PD) is used to control the position. Stiff actuators cannot store and release energy or exploit natural dynamics. They are energy inefficient, cannot exploit the natural dynamics of the system as found in passive walkers [20], have difficulties with explosive motions like throwing, jumping or kicking a ball since the maximum speed is limited and the high reflected inertia of the high-geared motor, makes them unsafe in case of impacts and clamping and as such they cannot absorb impacts occurring typically during jumping and running. One of the main differences between classical robots and humans is the presence of adaptive compliance in the actuated joints. This is largely due to the dynamic mechanical properties of the biological actuators: the muscle-tendon units. For example the work of Alexander explains the important role of biological springs in animals and humans during locomotion. The introduction of compliant elements in the robotic hardware started long ago with the use of pneumatic actuators and got well known with Raibert's work using air springs for hopping robots, later Pratt introduced the Series Elastic Actuators (SEA) [22]. For all the novel requirements of safety and energy-efficiency and applications the term Soft Robotics has been introduced. An important step was to make the stiffness of the compliant element adjustable. Variable impedance actuation (VIA) permits the embodiment of natural characteristics, found in biological systems into a new generation of mechatronic systems. A global overview of the state of the art of compliant actuators can be found in [23,24]. BioMot is producing a novel variable impedance actuators to generate a diverse repertoire of efficient skills. One of our key challenges is to study and build actuators with bi-directional torque and adaptable stiffness that are compact and implemented in assistive WRs. This research aims to determine the feasib ility and the advantages of Multiple DOF VIA and to reduce their size to a level that they can be implemented in a WR. Ultimately, our intention is to progress in the field of VIA by developing bi-articular actuators with adaptable compliance to allow for direct energy exchange between different lower limb joints.

2.5 Neuromusculoskeletal Modeling

Understanding human movement implies decoding the complex dynamics of the neural drive to muscles. This is the result of discrete events, generated in the brain, spinal cord, and nerves that ultimately determine the excitation of multiple muscles and the subsequent production of force and joint motion. Deciphering this process allows unravelling how the nervous and the musculoskeletal system are interfaced with one another and how the resulting neuromuscular mechanisms contribute to the generation of human movement. To understand the nature of these mechanisms it is first of all necessary to collect an experimental measurement of the neural drive sent to the muscles. Then, it is necessary to couple these measurements to physiologically accurate computational models of the human musculoskeletal system. This will allow modeling how neural inputs are translated into mechanical outputs by muscles and joints. We will refer to this process as neuromusculoskeletal (NMS) modeling. Recordings of surface electromyography (EMG) signals indirectly reflect the neural drive to muscles and can be easily recorded in vivo in the intact human subject during dynamic motor tasks [25]. Neuromusculoskeletal modeling based on surface EMG recordings (i.e. EMG-driven modeling) has demonstrated to be an effective way of predicting the actual behavior of muscles both on healthy [26] and unhealthy subjects [27]. This led to the possibility of determining the subject-specific relationship between patterns of muscle excitation and the resulting muscle dynamics without making any a-priori assumption on how muscles activate [26]. This is a main advantage with respect to previously proposed optimization-based methodologies and it allowed modeling and characterizing the form and function of muscles in subjects with neurological and orthopaedic conditions as well as different levels of muscle atrophy. This opened up to the possibility of studying how neuromuscular impairments contributed to abnormal movement and addressing motor limitations in subjects affected by different conditions including anterior cruciate ligament rupture, patellofemoral pain, osteoarthritis, stroke, and upper extremity neuromuscular injuries. BioMot investigates novel methods for adequate modulation of joint stiffness during walking, decreasing muscular effort needed to walk while preserving normal kinematics, with compliant, lightweight WRs. The optimum balance between the muscle joint torque and exoskeleton joint torque in terms of overall energy consumption is investigated with simulations. We aim to provide an approach to determine optimal features to understand variants in gait dynamics for control of active gait exoskeletons and to establish control methods for adaptable compliance in dynamic human-robot interaction.

2.6 Learning Mechanisms for Adaptive Walking

All living organisms have the capability of adapting to unknown environmental situations. Without using any supervising signals, they can create the behaviours that can survive in the natural environment. Recent advances in artificial learning and adaptive methods for robot control have not reached the level of adaptability of biological systems. One of the most critical problems with conventional approaches is the way of specifying goals of learning. The goals of learning in many cases are specified in advance by using supervising signals such as teaching signals in neural networks, cost functions in genetic algorithm, and reward function in reinforcement learning, which are not changed during learning even though their environments have changed. It would be difficult to adapt to environmental changes in learning that were supervised by fixed signals. Recent advances for artificial intelligence and cognitions in the field of human-robot interactions focus on the human's development and learning systems [28]. Kuniyoshi et al. [29] constructed a fetus model with sensory-motor brain model, and showed that various meaningful behaviors emerge through interactions with the environment in the womb. The notion of tacit learning has been proposed to develop artificial control systems with the adaptability to unpredictable environmental changes based on the control principle of biological systems [30]. The fundamental computational algorithm of tacit learning uses the feature of biological regulatory systems in which all regulations result from the spatial and temporal integration of homogeneous computational media that act subject to innate rules. A network of the homogeneous computational media that connects the sensors and motors in proper ways is of great advantage in orchestrating the flow of heterogeneous environmental information. Consortium partner RIKEN so far developed the networks of the artificial computational media, and experimentally showed that the gait of the 36DOF humanoid robot matches much better to the environment than the walking gaits of full-controlled humanoid robots and share some features with humans' natural gait in terms of the walking efficiency, adaptation of walking rhythm and the robustness on the walking terrain. These similarities of walking distinguish tacit learning from other learning architectures for creating bipedal walking [31]. The environmental information taken into the network through the reflexive actions played the roles of supervising signals for tacit learning. This learning scheme is strongly associated with the notion of affordance, which is recognized as the key factor in cognition and intelligence. In our case, the environmental information is mainly used to create the motions of the joints without concrete references and leads to the adapted behaviours. The creation of meaningful behaviours from purposeless actions by using the environmental information should be an essential process to establish adaptation and intelligence in man-made machines. This feature of tacit learning can accomplish high levels of adaptability and artificial intelligence. Our expected outcome is a new high-level learning strategy that employes abstraction of comfortable biomechanical configurations to adapt support based on experience. In this WP, the project aims at demonstrating the feasibility of a learning strategy that will not consider pre-defined variables but an abstraction of the perceptual responses that are directly related to a comfortable and safe support.

3   Summary

BioMot conducts experiments to reveal physical and cognitive adaptations between human-users and WRs. The main application scenarios that are considered to test these adaptations are (a) gait training for incomplete SCI users and (b) gait support for healthy users. The scientific production in this European consortium can be summarized in the following areas:

- Wearable exoskeletons for locomotion. Biomot will lead significant advancements in the field of exoskeleton robotic technologies. The robotic systems are equipped with artificial control systems that rely on physiological mechanisms to solve interactions with the environment. The required intelligence to perform stable locomotion will be founded on distributed cognitive and compliant skills for control of interaction during a number of situations. The BioMot project will produce a completely new approach for bioinspired control human-exoskeleton interaction. The BioMot WRs will allow the community to study and reveal basic principles governing human locomotion, motor recovery and aid gait rehabilitation after neurological and physical injury.
- Cognitive and behavioural contributions to gait dynamics. The cognitive mechanisms related to self-adjustments during walking (i.e. to detect the intention of initiation/termination of the gait, to detect the intention of changes in direction/orientation and to detect the intention of changing speed) are under experimental analysis with healthy and SCI patients in the project. In addition, the cognitive attention mechanisms contributing to stability and adaptation to environment during walking are analysed. Furthermore, brain decoders are under development to obtain the information of locomotion during walking from EEG signals. From the scientific point of view, these developments will have an outstanding impact in the fields of rehabilitation of lower limbs, brain-machine interfaces, and neuroscience, mainly.
- Compliant actuation of wearable robots for gait. Compliant actuators become more and more important as actuators for wearable assistive robotic devices for gait. They can reduce the size and weight of the actuators, have a natural, intrinsic compliant, interaction with the human and have intrinsic safety aspects. We are producing more sophisticated adaptable compliant actuation principles that could be used not only in wearable robots for gait, but also in rehabilitation robots, service robots and industrial robots.
- Neuromusculoskeletal modeling. Physiologically accurate, subject-specific, computational models of the human neuromusculoskeletal system are under development and testing in the BioMot project. These allow predicting how experimental measurements of neural inputs are translated into mechanical outputs by muscles and joints in real-time as an individual user moves. Real-time NMS modeling is also tunned to estimate how human subjects interact with the WR during dynamic motor tasks. In this context, methodologies are developed to estimate how human subjects continuously modulate their muscular and articular properties as a function of the basic constraints within the scenario. These developments are expected to provide the communitiy with an advanced human-machine interface that allows establishing more symbiotic relationships between the human-users and any WR to ultimately improve the level of co-adaptation and human-robot interaction.

References

1. Pons, J.L.: Wearable Robots: Biomechatronic Exoskeletons. Wiley, Chichester (2008)
2. Dollar, A.M., Herr, H.: Lower extremity exoskeletons and active orthoses: challenges and state-of-the-art. J. Mol. Biol. IEEE Trans. Robot. **24**(1), 144–158 (2008)
3. Taga, G.: A model of the neuro-musculo-skeletal system for human locomotion - I. Emergence of basic gait. Biol. Cybern. **73**(2), 97–111 (1995)
4. Lewis, C.L., Ferris, D.P.: Invariant hip moment pattern while walking with a robotic hip exoskeleton. J. Biomech. **44**(5), 789–793 (2011)
5. Stephen, C., Keith, G., Daniel, F.: Locomotor adaptation to a powered ankle-foot orthosis depends on control method. J. NeuroEng. Rehabil. **4**, 48 (2007)
6. Aguirre-Ollinger, G., Colgate, J.E., Peshkin, M.A., Goswami, A.: Design of an active one-degree-of-freedom lower-limb exoskeleton with inertia compensation. Int. J. Robot. Res. **30**(4), 486–499 (2011)
7. Esquenazi, A., Talaty, M., Packel, A., Saulino, M.: The ReWalk powered exoskeleton to restore ambulatory function to individuals with thoracic-level motor-complete spinal cord injury. Am. J. Phys. Med. Rehabil./Assoc. Acad. Physiatrists **91**(11), 911–921 (2012)
8. Ramachandran, P.: From science fiction to reality: exoskeletons. Life in Action **1**(3), 20–21 (2011)
9. Ivanenko, Y.P., Poppele, R.E., Lacquaniti, F.: Motor control programs and walking. Neuroscientist **12**(4), 339–348 (2006)
10. McGowan, C.P., Neptune, R.R., Clark, D.J., Kautz, S.A.: Modular control of human walking: adaptations to altered mechanical demands. J. Biomech. **43**(3), 412–419 (2010)
11. Ivanenko, Y.P., Grasso, R., Zago, M., Molinari, M., Scivoletto, G., Castellano, V., Macellari, V., Lacquaniti, F.: Temporal components of the motor patterns expressed by the human spinal cord reflect foot kinematics. J. Neurophysiol. **90**(5), 3555–3565 (2003)
12. Cappellini, G., Ivanenko, Y.P., Poppele, R.E., Lacquaniti, F.: Motor patterns in human walking and running. J. Neurophysiol. **95**(6), 3426–3437 (2006)
13. Ivanenko, Y.P., Cappellini, G., Poppele, R.E., Lacquaniti, F.: Spatiotemporal organization of alpha-motoneuron activity in the human spinal cord during different gaits and gait transitions. Eur. J. Neurosci. **27**(12), 3351–3368 (2008)
14. Grillner, S.: Biological pattern generation: the cellular and computational logic of networks in motion. Neuron **52**(5), 751–766 (2006)
15. Fitzsimmons, N.A., Lebedev, M.A., Peikon, I.D., Nicolelis, M.A.L.: Extracting kinematic parameters for monkey bipedal walking from cortical neuronal ensemble activity. Front. Integr. Neurosci. **3**(3), 1–19 (2009)
16. Castermans, T., Duvinage, M.: Corticomuscular coherence revealed during treadmill walking: further evidence of supraspinal control in human locomotion. J. Physiol. **591**, 1407–1408 (2013)
17. Petersen, T.H., Willerslev-Olsen, M., Conway, B.A., Nielsen, J.B.: The motor cortex drives the muscles during walking in human subjects. J. Physiol. **590**, 2443–2452 (2012)
18. Severens, M., Nienhuis, B., Desain, P., Duysens, J.: Feasibility of measuring event related desyncronization with electroencephalography during walking. In: International Conference of the IEEE EMBS, pp. 2764–2767 (2012)

19. Presacco, A., Goodman, R., Forrester, L., Contreras-Vidal, J.L.: Neural decoding of treadmill walking from noninvasive electroencephalographic signals. J. Neurophysiol. **160**(4), 1875–1887 (2011)

20. Collins, S.H., Ruina, A., Tedrake, R., Wisse, M.: Efficient bipedal robots based on passive-dynamic walkers. Science **18**(307), 1082–1085 (2005)

21. Alexander, R.: Three uses of springs in legged locomotion. Int. J. Robot. Res. (Special Issue on Legged Locomotion) **9**(2), 53–61 (1990)

22. Pratt, G.A., Williamson, M.M.: Series elastic actuators. In: IEEE International Workshop on Intelligent Robots and Systems (IROS 1995), Pittsburg, USA, pp. 399–406 (1995)

23. Van Ham, R., Sugar, T.G., Vanderborght, B., et al.: Compliant actuator designs review of actuators with passive adjustable compliance/controllable stiffness for robotic applications. IEEE Robot. Autom. Mag. **16**(3), 81–94 (2009)

24. Vanderborght, B., Albu-Schaeffer, A., Bicchi, A., et al.: Variable impedance actuators: a review. Robot. Auton. Syst. **61**(12), 1601–1614 (2013)

25. Farina, D., Negro, F.: Accessing the neural drive to muscle and translation to neurorehabilitation technologies. IEEE Rev. Biomed. Eng. **5**, 3–14 (2012)

26. Sartori, M., Reggiani, M., Farina, D., Lloyd, D.G.: EMG-driven forward-dynamic estimation of muscle force and joint moment about multiple degrees of freedom in the human lower extremity. PLoS ONE **7**, 12 (2012)

27. Higginson, J.S., Ramsay, J.W., Buchanan, T.S.: Hybrid models of the neuromusculoskeletal system improve subject-specificity. Proc. Inst. Mech. Eng. Part H: J. Eng. Med. **226**(2), 113–119 (2011)

28. Asada, M., Hosoda, K., Kuniyoshi, Y., Ishiguro, H., Inui, T., Yoshikawa, Y., Ogino, M., Yoshida, C.: Cognitive developmental robotics: a survey. IEEE Trans. Auton. Ment. Dev. **1**(1), 12–34 (2009)

29. Kuniyoshi, Y., Sangawa, S.: Early motor development from partially ordered neural-body dynamics: experiments with a cortico-spinal-musculo-skeletal model. Biol. Cybern. **95**, 589–605 (2006)

30. Shimoda, S., Kimura, H.: Bio-mimetic approach to tacit learning based on compound control. IEEE Trans. Syst. Man Cybern.-Part B. **40**, 77–90 (2010)

31. Matsubara, K., Morimoto, J., Nakanishi, J., Sato, M., Doya, K.: Learning CPG-based biped locomotion with a policy gradient method. Robot. Auton. Syst. **54**(11), 911–920 (2006)

Experimenting with Users

Measuring User Acceptance of Wearable Symbiotic Devices: Validation Study Across Application Scenarios

Anna Spagnolli[✉], Enrico Guardigli, Valeria Orso,
Alessandra Varotto, and Luciano Gamberini

Department of General Psychology and HIT Research Center,
Via Venezia 8, 35131 Padua, Italy
{anna.spagnolli,luciano.gamberini}@unipd.it,
eguardigli@gmail.com, valeria.orso@studenti.unipd.it,
alevarotto@hotmail.com
http://htlab.psy.unipd.it/

Abstract. Wearable devices detecting users' psycho-physiological parameters and providing related feedback are an important component of intelligent systems adapting to users' cognitive and affective states. However, issues related to perceived comfort and privacy might compromise users' intention to use them in real contexts. To measure users' acceptance of these devices, we built a questionnaire that includes key dimensions of the TAM model [7, 15], such as perceived usefulness, effort expectancy, psychological attachment, facilitating conditions, and some dimensions that are especially relevant to wearable symbiotic systems (e.g., perceived comfort, and perceived privacy). This questionnaire was administered to 110 respondents with reference to three devices (i.e., smart -shirt, portable EEG system, and eye-tracking glasses) and six usage scenarios (dangerous work, heavy work, sport, homecare, research, retail).

After validation, 26 items were retained for the analysis and their factorial structure clarified. Perceived usefulness, perceived comfort/pleasantness, facilitating conditions, and attitude toward technology are good predictors of acceptance. The effects of scenario, device, and expertise are also discussed.

Keywords: Wearable computers · Symbiotic system · User acceptance

1 Introduction

Wearable computers are fully functional, self-contained electronic devices that can be worn, carried, or attached to the body, letting the user access information anytime and anywhere [1–3]. Unlike generic portable devices, wearable devices do not require muscular effort to be carried around, remain attached to the body regardless of its orientation and activity, and do not need to be removed from the body in order to be operated [4]. The ubiquitous, portable nature of wearable computers makes them an ideal component of symbiotic systems (i.e., systems that record and interpret a user's cognitive and affective states and respond accordingly). In particular, they can be used

© Springer International Publishing Switzerland 2014
G. Jacucci et al. (Eds.): Symbiotic 2014, LNCS 8820, pp. 87–98, 2014.
DOI: 10.1007/978-3-319-13500-7_7

to record psycho-physiological parameters such as heart rate, breathing rate, electro-dermal activity, eye activity, cerebral activity, and muscular activity [5]. In this way, the human-computer interface is relocated from an external device onto the body itself, which provides—almost inadvertently—input to the system [2].

Current models of user acceptance include factors such as perceived usefulness, or the extent to which using a system will increase performance; effort expectancy, or the ease of use associated with the device; psychological attachment, or the benefits derived from adopting a device to maintain some relations or comply with values; and facilitating conditions, or the extent to which the technical and organizational structure can support the use of a device [6, 7]. While these dimensions are of general relevance, we argue that comfort and privacy should also be considered while measuring acceptance of wearable symbiotic devices. Indeed, the drawback is that physical factors such as size, weight, and textile fibers of the device can negatively affect the level of comfort experienced by the user [2, 8] and then prevent their adoption in real contexts. Also, they might be perceived as funny or embarrassing to be worn in public [9]. Finally, the kind of data collected raises privacy issues because the information col-lected can lead to the identification of the user or be sensitive—that is, one that might result in loss of an advantage or level of security if disclosed to others who might have low or unknown levels of trust or undesirable intentions. Unpleasant consequences might include misinterpretation of data, disregarding their original meaning [10], or profiling (by merging databases or through data-mining) and attribution of new property to the individual derived from implicit patterns [11].

In the present study, we use a questionnaire to assess user acceptance of wearable symbiotic devices that includes all the abovementioned dimensions. The goal of the study is to extract the underlying factorial structure of the questionnaire and to define which dimensions are responsible for the intention to use those devices. To help respondents formulate opinions about technologies that might otherwise be too unusual or far from their everyday life, the questionnaire was contextualized into specific usage scenarios, and respondents were recruited among people belonging to those scenarios, in addition to generic users. The rest of this paper describes the method and the results of the study. The last section discusses the results.

2 Questionnaire

The section of interest of the questionnaire consists of 45 items, preceded by a brief section collecting background information and screening questions to ensure that each respondent effectively belonged to the expected subgroup. The items are meant to measure the following dimensions:

1. Attitude Toward Technology (ATT), 6 items. This regards an individual's overall reaction toward the use of technological devices [7]. The items refer to the survey developed by [12] to assess young students' attitudes toward technology, or the Pupils' Attitude Towards Technology (PATT) questionnaire (e.g., "Nowadays I think information technology is indispensable").

2. Technology Anxiety (TA), 5 items. Based on the concept of computer anxiety [13], it refers to the feeling of apprehension or even fear when using technology. The items investigating TA were adapted from [14] (e.g., "The possibility of using a device or a technology that I have never used before makes me feel anxious").

3. Facilitating Conditions (FC), 3 items. This dimension refers to the extent to which the user perceives that there are factors able to support the system deployment [7] (e.g., "The device would be incompatible with most aspects of my activity").

4. Perceived Usefulness (PU), 4 items. This dimension refers to the extent to which the user believes that the use of a certain device would enhance her performance [15] (e.g., "The device could improve my performance").

5. Effort Expectancy (EE), 2 items. This dimension refers to the perceived ease associated with the use of a certain device [7] (e.g., "It seems easy to learn how to use the device").

6. Behavioral Intentions (BI), 4 items. This dimension refers to the extent to which an individual has formulated conscious plans to carry out a certain action using a particular device [15] (e.g., "If the device was available to me, I would use it").

7. Psychological Attachment (PA), 5 items. This dimension refers to the individual's tendency to adopt a technology as a consequence of social influence [16]. The items investigating PA were adapted from the questions used by [16] (e.g., "If most people in my environment used the device, I would be more inclined to use it as well").

8. Perceived Enjoyment (PE), 4 items. This dimension refers to the extent to which users perceive the activity of using a system to be pleasant, aside from any consequences resulting from device use [13] (e.g., "I think the device was boring").

9. Perceived Comfort (PC), 7 items. This dimension refers to individuals' perceived comfort in wearing a wearable computer. The items were selected and adapted from the Comfort Rating Scale developed by [17] (e.g., "I think the device is well suited to my body").

10. Perceived Privacy (PP), 5 items. This dimension refers to the extent to which the user is confident that the data recorded by the device is safely handled and stored. The items were selected and adapted from a questionnaire developed by [18] to assess privacy concerns on the use of health information (e.g., "I think that the device threatens my privacy").

Dimensions 1–8 were derived from the Technology Acceptance Model and Unified Theory of Acceptance and Use of Technology [7, 15]; dimensions 9–10 were adapted from [17, 18].

Each item was formulated as a statement. Participants were asked to rate the extent to which they agreed or disagreed with each statement on a 6-point Likert scale (1 = "completely disagree") for each of the three devices under assessment (except for ATT, TA, and PP, where statements regarded technology in general and not specific devices). A small icon represented the device on the response scale. The option "I don't know" was also included in each scale.

In addition to the items above, we also asked participants to specify to whom they would have disclosed the data collected by the devices.

3   Method

3.1   Sample

A total of 110 participants (33 women) volunteered in the present research (mean age 32.5, $SD = 13.77$). Each participant either belonged to one target category (experts) or was a generic user (nonexpert). The target categories were selected to fit the different scenarios in which the questionnaire was contextualized and included firemen for the dangerous job scenario, factory workers for the physical demanding job scenario, volleyball players for the sport scenario, elderly people for the homecare scenario, and neuroscientists for the research scenario. All scenarios but retail were presented to 10 experts and 10 university students. In the retail scenario, only a group of 10 university students was used.[1]

3.2   Setting and Procedure

Except for the student subgroup, the respondents were met on premises agreed upon. Respondents were first told briefly the nature of the study; then they read and signed the informed consent. Afterward, a video described the main functionalities of the three devices under assessment (i.e., the smart t-shirt, the portable EEG system, the eye-tracking glasses), and participants were asked to wear the devices. After this demonstration phase, one of six possible scenarios was presented (sport, physical demanding job, dangerous job, homecare, retail, and research) via short narratives. Below is a report of the narrative used for the research scenario:

"You and your colleagues are asked to use some wearable devices, a smart t-shirt with heart rate and breath rate monitor, a helmet with EEG and a pair of glasses for tracking the gaze, during your daily working activities. These psychophysiological monitoring devices would make visible some of your unconscious processes while you are performing different activities and as a response to specific stimuli. For example they might be useful to identify relevant data for the researcher when she is handling large datasets."

Once the narrative was read, the participant was administered the questionnaire. The entire session took 30–40 min.

3.3   Analysis

A principal component analysis (PCA) with orthogonal rotation (VARIMAX) was conducted on the initial 43 items measuring user acceptance. For items with separate responses for each device, the average score between the three devices was considered. The aim of such analysis was to reduce the number of the items of the questionnaire by eliminating those poorly correlating with the others in the same cluster. As a result, 19 items were removed from the original list. A second principal component analysis was

---

[1] The retail scenario is a very common one in everyday life, so general users can hardly be considered as non-experts in it.

then run to test the validity of the model composed by the 26 items left. Preliminarily, the Kaiser-Meyer-Olkin index verified the sampling adequacy for the analysis, KMO = .725. Also, Bartlett's test of sphericity χ^2 325 = 1190.35, p < .001, indicated that correlations between items were sufficiently large for PCA. The analysis was run to extract 10 components. The 10 components extracted all had Eigenvalues over Kaiser's criterion of 1 and in combination explained the 75.53 % of the variance, which is above the cut-off value of 70 %.

Regression models were run with behavioral intention as a dependent variable, and the dimensions were assessed by the questionnaire together with the demographic values of age, gender, and education as hypothesized predictors.

All analyses were run with SPSS v. 20.

3.4 Hypotheses and Research Questions

The study meant first to identify the factorial structure that can explain the variability of the answers and to simplify the questionnaire accordingly. On the simplified questionnaire, the main hypotheses regarded the role of the two main factors responsible for behavioral intention according to the TAM model: perceived usefulness, effort expectancy, the effect of the dimensions (especially those we introduced in the questionnaire), perceived comfort, technology anxiety, and attitude towards technology.

H1a: Perceived Usefulness is a significant predictor of the intention of using a device.

H1b: Effort Expectancy is a significant predictor of the intention of using device.

H2: Perceived Comfort is a significant predictor of the intention of using a device.

H3a: Technology Anxiety is a significant predictor of the intention of using a device.

H3b: Attitude toward Technology is a significant predictor of the intention of using a device.

Finally, we hypothesized the effect of expertise on acceptance in the following way:

H4: Expert users will have higher acceptance scores compared to nonexpert users.

We also explored the effect of the scenarios on the acceptance level and the extent to which privacy affects the intention of using wearable symbiotic devices; in this regard, we will also consider how respondents are willing to disclose their personal information to different types of hypothetical collectors.

4 Results

4.1 Questionnaire Validation

Table 1 shows the factor loadings after rotation. The items that cluster on the same component suggest that component represents perceived comfort (PC), component 2 perceived usefulness (PU), component 3 behavioral intention (BI), component 4 technology anxiety (TA), component 5 attitude toward technology (ATT), component

6 effort expectancy (EE), component 7 facilitating conditions (FC), component 9 psychological attachment (PA), and component 10 perceived privacy (PP). Items that were supposed to measure perceived enjoyment (PE) seem to load on components 1 and 8. This might be attributed to an overlap in the two dimensions of comfort and enjoyment, which are both associated with a positive or negative experience with the device. However, comfort is more connected to the physical aspect of the experience as evidenced by Item 2, which could not be categorized as PC, so we decided not to collapse the two factors into one. Moreover, the limited amount of time during which the respondents have used the devices might also account for this overlap because devices could not differentiate much in terms of fatigue after prolonged usage. Therefore, items hypothetically belonging to PC are considered to cluster on component 1 (items 24, 25, and 26) while those intended to capture PE are considered to cluster on component 8 (items 21, 22, and 23).

4.2   Predictors of Intention to Use the Device

The most satisfactory linear regression model included four dimensions assessed by the questionnaire (ATT, FC, PU, PE), three scenarios of use (dangerous job, healthcare, and neuroscience), and age (Table 2). In this model, PU and PE were found to be good predictors of behavioral intention, respectively: $\beta = .355$, $p < .001$; and $\beta = .290$, $p < .001$. FC was found to have a moderate influence in determining score of BI, $\beta = .17$ $p = .043$. Similarly, ATT was a moderately strong predictor: $\beta = .15$ $p = .039$. Dangerous job, homecare, and neuroscience scenarios were predictors: $\beta = .17$, $p = .025$; $\beta = .23$, $p = .01$; and $\beta = .18$. $p = .02$. Age was found to negatively predict behavioral intention —$\beta = -.308$, $p < .001$—indicating that the older the respondents, the lower the acceptance. The abovementioned model explained the 47 % variance in BI.

Finally, respondents were asked to choose with whom they would be likely to share the data recorded by the wearable devices. They could choose among physician, psychologist, partner, relatives, friends, research centers, personal trainer, colleagues, police, commercial services, boss, shopping assistant, or no one. To disclose data about one's emotions, only psychologist and partner were selected by more than 50 % of the

Table 1. Factor Loadings after rotation with the 26 items selected after the first principal components analysis. Values reported in bold indicate items loaded on that particular component.

	C1	C 2	C3	C 4	C5	C6	C7	C8	C9	C10
	P C	PU	BI	TA	ATT	EE	FC	PE	PA	PP
1. Overall I think information technology brings about some benefits.	0.124	-0.03	0.345	0.166	**0.658**	0.123	-0.036	-0.173	-0.166	0.222
2. I think information technology is indispensable today.	0.033	0.274	-0.12	0.104	**0.787**	-0.09	0.164	0.025	0.077	0.06

(Continued)

Table 1. (*Continued*)

| | C1 | C 2 | C3 | C 4 | C5 | C6 | C7 | C8 | C9 | C10 |
	P C	**PU**	**BI**	**TA**	**ATT**	**EE**	**FC**	**PE**	**PA**	**PP**
3. I constantly have to deal with information technology.	-0.115	-0.02	0,105	0.022	**0.786**	0.097	-0.036	0.144	0.073	-0.069
4. When I have to use some information and communication technology, I fear that I will break it or make irreversible mistakes.	-0.032	-0.17	0.104	**0.696**	0.007	0.09	0.377	-0.058	0.324	0.069
5. Most issues connected to technology are difficult for me.	0.038	-0.17	0.082	**0.868**	0.101	0.086	0.064	-0.053	-0.119	0.025
6. The possibility of using a device or a technology that I have never used makes me feel anxious.	0.009	0.059	-0.01	**0.874**	0.083	0.046	0.037	0.158	-0.035	0.004
7. The device would be incompatible with most aspects of my activity.	0,084	0,083	0,153	0,174	0,078	0,033	**0,844**	0,006	0,076	-0,042
8. The device limits the way in which I like to perform my activity.	0.246	0.277	0.211	0.15	-0.005	0.131	**0,679**	0.209	-0.076	-0.202
9. The device could help in reaching my objectives.	0.082	**0.75**	0.375	-0.087	-0.032	0.092	0.212	0.004	0.018	0.161
10. The device could improve my performance.	0.205	**0.839**	0.065	-0.072	0.162	0.009	-0.055	0.015	0.143	-0.066
11. The device could improve the quality of my activity.	-0.017	**0.757**	0.314	-0.122	0.075	0.108	0.22	-0.021	0.233	-0.06
12. It seems easy to learn how to use the device.	0.06	0.064	0.064	0.14	0.038	**0.852**	0.023	-0.052	-0.019	-0.055
13. It seems tiresome to use the device.	0.188	0.079	0.08	0.111	0.06	**0.631**	0.143	0.418	0.02	-0.072
14. If the device were available to me, I would use it.	0.259	0.141	**0.697**	-0.168	0.185	0.107	0.227	0.053	0.266	0.01
15. If it were launched on the market at an affordable price, I would likely purchase it.	0.066	0.247	**0.811**	0.104	-0.024	0.023	0.1	0.027	0.082	-0.044
16. I think I would use the device only if I were forced to.	0.084	0.202	**0.706**	0.158	0.125	0.031	0.097	0.278	-0.065	-0.128

(*Continued*)

Table 1. (*Continued*)

	C1 P C	C 2 PU	C3 BI	C 4 TA	C5 ATT	C6 EE	C7 FC	C8 PE	C9 PA	C10 PP
17. If people who are influential in my life recommended that I use the device for a period of time, I would do so.	0.333	0.078	0.197	-0.225	0.294	0.24	0.043	-0.008	**0.607**	0.077
18. If most people in my environment used the device, I would be more inclined to use it as well.	-0.101	0.334	0.058	0.107	-0.067	-0.16	0.013	0.039	**0.789**	-0.083
19. I think that privacy breaches are a serious issue today.	-0.049	0	-0.05	0.13	-0.014	-0.24	-0.075	0.038	0.003	**0.833**
20. I think that the device threatens my privacy.	0.219	-0.01	-0.1	-0.099	0.192	0.384	-0.099	0.103	-0.051	**0.62**
21. I think the device was pleasant	**0.633**	0.25	0.32	0.085	-0.227	-0.04	-0.045	**0.147**	-0.065	-0.054
22. I think using the device was annoying.	**0.713**	0.009	0.071	0.061	0.015	0.168	0.101	**0.49**	0.008	0.091
23. I think the device was boring.	**0.362**	-0.07	0.249	0.091	0.06	0.024	0.012	**0.745**	0.059	0.069
24. In think the device was comfortable.	**0.796**	0.165	0.111	-0.029	-0.048	0.065	0.111	0.163	-0.02	0.003
25. I think the device is well suited to my body.	**0.674**	-0.12	-0.05	-0.031	0.363	0.258	0.228	-0.062	0.215	0.124
26. Wearing the components feels weird physically.	**0.301**	0.105	0.043	-0.133	0.032	0.414	0.297	0.49	-0.074	0.286

sample. Respondents would share information about their level of stress, mental states, and cognitive performance with physician, psychologist, and partner. Interest and preferences were preferably shared with friends and partner.

4.3 Effect of Device and Expertise

A repeated measures ANOVA highlighted a significant main effect of device and expertise on behavioral intention, respectively: $F_{2,107} = 22.63$, $p < .001$, $\eta_p^2 = .29$; and $F_{1,107} = .7.65$, p = .007, $\eta_p^2 = .06$. Regarding the effect of the device, post hoc comparisons showed that the smart t-shirt ($M = 4.24$, $SD = .12$) received significantly more positive evaluations compared to both the portable EEG ($M = 3.37$, $SD = .13$) and the eye-tracking glasses ($M = 3.22$, $SD = .13$). No significant difference emerged between the scores of the portable EEG and the eye-tracking glasses. Concerning the effect of the expertise, post hoc comparisons also highlighted a significant difference in the way experts and nonexperts gave their scores: nonexpert users ($M = 3.91$, $SD = .16$) gave more positive evaluations compared to experts ($M = 3.31$, $SD = .14$). All post hoc

Table 2. Linear regression model. In bold are the statistically significant coefficients. Significance level * = $p < .05$ ** = $p < .001$; $N = 110$.

Linear Regression Model			
	Standardized Coefficients		
Model	Beta	t	p
(Constants)		-.262	.794
ATT	.154	2.093	.039*
FC	.168	2.050	.043*
PU	.355	4.667	,000**
PE	.290	3.881	.000**
Dangerous Job	.174	2.277	.025*
Homecare	.229	2.642	.010*
Neuroscience	.182	2.373	.020*
Age	-.308	-3.803	.000**

comparisons were run with Bonferroni correction (Table 3). The interaction of device and expertise was not significant.

For the effect of expertise on PP, ATT, and TA (when the device did not vary), one-way ANOVAs found a significant effect of expertise for PP privacy ($F_{1,108} = 5.94$, $p = .016$), and nonexperts seemed more concerned about privacy than experts (respectively, $M = 2.77$, $SD = 1.24$; and $M = 3.35$, $SD = 1.23$).

Table 3. Repeated measures ANOVA for Device*Expertise, referring to Behavioral Intention. Significance level * = $p < .05$ ** = $p < .001$

		M(SD)	Test	η^2_p
Device	Smart t-shirt	4.24 (± .12)		
	Portable EEG	3.37 (± .13)	$F_{2, 107} = 22.63**$	.29
	Eye-tracking glasses	3.22 (± .13)		
Expertise	Expert	3.31 (± .14)	$F_{1, 107} = .7.65*$	.066
	Nonexpert	3.91 (± .16)		
Device*Expertise			$F_{2, 218} = .75$	.014

5   Discussion

Consistent with the classic models of user acceptance [7, 13], PU and PE were found to play a relevant role in determining the level of acceptance of a device. Contrary to our hypothesis, EE was not a significant predictor of user acceptance probably because participants did not use the devices during a prolonged activity but instead used them only for a few minutes. Of the other dimensions introduced in the questionnaire, PC seemed to overlap with classic PE; ATT (but not TA) predicted users' acceptance rates. Although PP did not seem to significantly affect acceptance, the open question suggested that respondents seem quite selective in defining which data would be disclosed to whom, unveiling concerns and differentiations that the items measuring PP in general could have masked. This points to the need to carefully address the specific meaning, preferences, and risks attributed to privacy by the target recipients of a symbiotic device. The kind of privacy loss that might be perceived as unacceptable to some categories of users might seem acceptable to others. For example, in our study, experts seemed to have less privacy concerns than nonexperts, probably because they are more used to releasing some personal information for the sake of their activity. Indeed, the scenarios that most affected acceptance were dangerous job, homecare, and neuroscience, where the users' dependence and even intimacy with technology are forcedly higher than in other scenarios. That wearable devices are positively accepted in medical contexts is also consistent with previous studies (e.g., [19–21]).

The other result—that experts considered the devices to be less useful than nonexperts—might depend on their reluctance to adopt and learn a new technology so different from the ones they already use. For the same and other reasons, older users seemed less inclined to use wearable devices, as already highlighted by [19]. Regarding devices, the favorable opinions about the smart t-shirt compared with portable EEG and the eye-tracking glasses is in line with [8], suggesting that lighter and more discrete wearable devices are better appreciated compared to bulkier and more noticeable ones.

6   Conclusions

The main contribution of this paper is the validation of a questionnaire on user acceptance of devices that represent a well-centered instance of symbiotic systems, which are characterized—among other features—by the advanced sensing capability of personal data and by dynamically adapting the output to the sensed data [22]. The analysis singled out the most robust items across different application scenarios by using classic dimensions typical of TAM models as well as new dimensions that seemed especially relevant to symbiotic systems (i.e., PC of the sensing device, PP, and ATT). Thus, in addition to offering a validated tool to measure acceptance of symbiotic systems, the study also has relevance to theories of technology acceptance in that it tests and enriches the classic model.

Acknowledgements. This research was supported by the European Project CEEDS (N: 258749; Call: ICT 2009.8.4 Human - Computer Confluence). The authors would also like to thank the participants and the following organizations: Associazione "La casa azzurra"; Centro sociale per anziani età d'oro; Comando provinciale dei vigili del fuoco di Padova; Siderurgica Ravennate Srl; Dipartimento Psicologia Generale, Università di Padova. The authors are also grateful to Enrico Tonini for advising about the statistic analysis.

References

1. Buenaflor, C.B., Kim, H.C.K.: Six human factors to acceptability of wearable computers. Int. J. Multimed. Ubiquitous Eng. **8**, 103–114 (2013)
2. Barfield, W., Mann, S., Baird, K., Gemperle, F., Kasabach, C., Stivoric, J., Cho, G.: Computational Clothing and Accessories. Fundamentals of Wearable Computers and Augmented Reality. Lawrence Erlbaum Associates, Mahwah (2001)
3. Duval, S., Hoareau, C., Hashizume, H.: Humanistic needs as seeds. In: Wong, W.K., Guo, Z.X. (eds.) Smart Clothing, pp. 153–187. Taylor and Francis Group, Boca Raton (2014)
4. Knight, J.F., Deen-Williams, D., Arvanitis, T.N., Baber, C., Sotiriou, S., Anastopoulou, S., Gargalakos, M.: Assessing the wearability of wearable computers. In: 10th IEEE International Symposium on Wearable Computers, pp. 75–82. IEEE Computer Society (2006)
5. Allanson, J., Fairclough, S.H.: A research agenda for physiological computing. Interact. Comput. **16**, 857–878 (2004)
6. Dillon, A.: User acceptance of information technology. In: Karwowski, W. (ed.) Encyclopedia of Human Factors and Ergonomics. Taylor and Francis, London (2001)
7. Venkatesh, V., Morris, M.G., Davis, G.B., Davis, F.D.: User acceptance of information technology: toward a unified view. MIS Q. **27**, 425–478 (2003)
8. Knight, J.F., Baber, C.: A tool to assess the comfort of wearable computers. Hum. Factors J. Hum. Factors Ergon. Soc. **47**, 77–91 (2005)
9. Rico, J., Brewster, S: Usable gestures for mobile interfaces: evaluating social acceptability. In: Proceedings of the 28th International Conference on Human Factors in Computing Systems (CHI '10), pp. 887–896. ACM, New York (2010)
10. Britz, J.: Technology as a threat to privacy: ethical challenges and guidelines for the information professionals. Microcomput. Inf. Manage. **13**, 175–193 (1996)
11. Tavani, H.T.: Genomic research and data-mining technology: implications for personal privacy and informed consent. Ethics Inf. Technol. **6**, 15–28 (2004)
12. DeVries, M.: Pupils' Attitude Towards Technology (PATT) - USA Instrument (1988). http://www.iteaconnect.org/Conference/PATT/PATTSI/PATTSurveyInstrument.pdf. Accessed February 2014
13. Venkatesh, V., Davis, F.D.: A theoretical extension of the technology acceptance model: four longitudinal field studies. Manage. Sci. **46**, 186–204 (2000)
14. Meuter, M.L., Ostrom, A.L., Bitner, M.J., Roundtree, R.: The influence of technology anxiety on consumer use and experiences with self-service technologies. J. Bus. Res. **56**, 899–906 (2003)
15. Davis, F.D.: A technology acceptance model for empirically testing new end-user information systems: theory and results. Doctoral dissertation, Massachusetts Institute of Technology (1985)
16. Malhotra, Y., Galletta, D. F.: Extending the technology acceptance model to account for social influence: theoretical bases and empirical validation. In: Proceedings of the 32nd Annual Hawaii International Conference on Systems Sciences, HICSS-32. p. 14. IEEE Computer Society (1999)
17. Kaplan, S., Okur, A.: The meaning and importance of clothing comfort: a case study for Turkey. J. Sens. Stud. **23**, 688–706 (2008)
18. Perera, G., Holbrook, A., Thabane, L., Foster, G., Willison, D.J.: Views on health information sharing and privacy from primary care practices using electronic medical records. Int. J. Med. Inf. **80**, 94–101 (2011)

19. Ziefle, M., Rocker, C.: Acceptance of pervasive healthcare systems: a comparison of different implementation concepts. In: 4th International Conference on Pervasive Computing Technologies for Healthcare (PervasiveHealth), pp. 1–6. IEEE Computer Society (2010)

20. Bergmann, J.H.M., McGregor, A.H.: Body-worn sensor design: what do patients and clinicians want? Ann. Biomed. Eng. **39**, 2299–2312 (2011)

21. Bodine, K., Gemperle, F.: Effects of functionality on perceived comfort of wearables. In 16th International Symposium on Wearable Computers, pp. 57–57. IEEE Computer Society (2003)

22. Jacucci, G., Spagnolli, A., Freeman, J., Gamberini, L.: Symbiotic interaction: a critical definition and comparison to other human-computer paradigms. In: Proceedings of Symbiotic 2014: Third International Workshop on Symbiotic Interaction. Springer, London (2014)

How Semantic Processing of Words Evokes Changes in Pupil

Patrik Pluchino[1]([✉]), Luciano Gamberini[1],
Oswald Barral[2], and Filippo Minelle[1]

[1] Human Inspired Technology Research Centre (HIT),
University of Padova, Padua, Italy
{patrik.pluchino,luciano.gamberini}@unipd.it,
minelle.filippo@studenti.unipd.it
[2] Helsinki Institute for Information Technology (HIIT),
Department of Computer Science, University of Helsinki, Helsinki, Finland
barralme@cs.helsinki.fi

Abstract. This paper investigates the relationship between semantic processing of words and modifications in pupil size. Variations in pupil diameter reflect cognitive processing, as has been widely demonstrated in literature. We designed an experiment in which semantic association between words was manipulated in order to disclose potential differences in cognitive processing. Moreover, we measured the concurrent pupil diameter changes. Results showed faster pupil dilation in trials in which words were semantically associated. As changes in pupil diameter do not occur under voluntary control, they could reflect processing of preconscious information. We believe that a better symbiotic relationship between humans and machines is achievable once systems are able to make us aware of these "involuntary" changes.

Keywords: Pupil diameter · Semantic association · Symbiotic relationship

1 Introduction

The pupillary system has been widely investigated for more than 50 years in psychology as an indicator of cognitive processing. It has been demonstrated that variation in pupil size are related to changes in mental states, in allocation of attention, and in consolidation of perceptive stimuli. The phenomenon of pupil dilation is linked to arousal, interest, and cognitive effort [1,2]. The pupil diameter can also be used to measure emotional states. For instance emotion (i.e. emotional and sexual arousal [2,3]), emotional content of words [4], and anticipation of knowing the answers to questions can increase pupil diameter [5].

The focus in the research field of cognitive psychology and psychophysiology is about the changes in pupil diameter related to processing of specific events. In literature, evidences were provided in order to support the relationship between mental elaboration (e.g. problem solving or language comprehension) and pupil

© Springer International Publishing Switzerland 2014
G. Jacucci et al. (Eds.): Symbiotic 2014, LNCS 8820, pp. 99–112, 2014.
DOI: 10.1007/978-3-319-13500-7_8

dilation [1]. Moreover, it has been demostrated how cognitive tasks that impose higher memory load are linked to an increment in pupil dilation [6]. Indeed, mental workload increases pupil diameter [7,8]. It has been revealed that increasing the difficulty of mental multiplication problems enlarges pupil diameter [9,10]. Moreover, it has been shown how different types of linguistic tasks (i.e. listening to, repeating back a message, or simultaneous interpretation of a foreign language message into the participant's native language) reflect variations in pupil size that are proportional to the level of mental effort [11]. Tasks in which participants have to read sentences of various syntactic complexity revealed changes in pupil diameter [12]. It has been proved that an increment in short-term memory load (i.e. 3 vs. 7 digits) increases pupillary dilation [13]. Futhermore, a higher cognitive workload during visual search tasks is mirrored by a proportional enlargement of the pupil [14]. Mean pupil dilation has been found to be a useful event-related index in order to measure cognitive load in research on learning [15]. Moreover, a study has demonstrated that lying produces pupil dilation [16,17] and the higher cognitive workload that is connected to lying (namely, attempts "to make lies as believable as possible") is more demading than a task of telling the truth [18]. Thus, reviewing the literature provides factual data regarding how cognitive workload is a fundamental concept in investigating both human cognition, to evaluate human-machine interfaces and to reach a better human-computer symbiotic interaction (for a theoretical overview and definition in the field of Human Computer Interaction see [19]). In 1960 [20], Licklider coined the term man-computer symbiosis. The idea concerned a mutual and constant "cooperative behaviour "between man and machine considering them as two dissimilar "organisms"in intimate association. Taking the concept of intimate association to the extreme it is possible to consider also wearable symbiotic devices (e.g. portable eye tracker and mobile EEG systems). Moreover, in order to enhance the symbiotic interaction it is fundamental to consider the degree of acceptance and the user experience of these tools during the interaction [21]. Furthermore, machines can provide useful information about changes in physiological indices (e.g. pupil dilation, galvanic skin response, heart rate) that are linked to cognitive processing of which human beings often are not conscious (for an overview, see [22–24]). It could be crucial to this special type of man-computer relationship that, for instance, machines could measure the level of cognitive workload in order to support, in some way, efficient human cognitive processing. The computers could reduce people's cognitive load by, for example, decreasing the amount of information displayed on the screen when they detect cognitive overload.

Cognitive psychophysiology has widely studied task-evoked pupillary responses (TEPRs); namely, the changes in pupil diameter that are linked to mental or sensory events. It is fundamental to consider how small these modifications are compared to variations in diameter of the pupil due to changes in light intensity. Following changes in illumination (e.g. from standard lighting conditions to dim lighting conditions), pupil diameter can even double its typical size. Instead, cognitively driven modifications in pupil size are usually on a smaller scale and are rarely more than 0.5 mm [25]. Thus, it is important to keep in mind that varying the brightness of the experimental stimuli that are presented on the screen may introduce artifacts in the recorded data [26]. It has to be taken into account that pupillary responses provide a continuous measure of cognitive processing despite the fact that a participant is not conscious of them. Thus, pupil can be seen as "a window to the preconscious" [27], considering that it can vary in conjunction with a low level of cognitive activation (i.e. a perceptive stimulus that is still not completely processed) that cannot reach the threshold of consciousness for a participant. On the one hand, pupillary responses usually occur spontaneously and are not under voluntary control despite the fact that there is proof that incremental pupil dilation can be provoked through mental sexual imagery. In this task, participants had to imagine an event or an object [28]. On the other hand, it is impossible to abolish a pupillary dilation that starts in response to external stimuli or mental events [29]. Regarding the technique employed to evaluate pupillary diameter changes, namely eye tracking (i.e. with a remote eye tracker), it is important to emphasise that this methodolgy is totally noninvasive. Indeed, the participant is seated in front of the screen at a fixed distance and the pupil can be measured without a chin rest. This is a strong advantage over other kinds of physiological measurements (e.g. EEG, fMRI) that present some forms of constraints related, for instance, to limited freedom of movement.

1.1   Semantic Associations

The semantic associations that are tested in the present study derive from the database of the Semantic Priming Project (SPP). It contains descriptive and behavioural measures for 1,661 target words and different types of associated words: first-associate related, first-associate unrelated, other-associate related, and other-associate unrelated [30]. An advantage of utilising this database is that it enables the realisation of better-controlled studies. Indeed, the paired words (i.e. prime–target) are generated with specific characteristics (e.g. length, frequency, associative strength, etc.).

The purpose of the current study is to underline possible differences in pupil dilation that are related to semantic associations. We expected the processing of semantically associated words to be different than the processing of words that are not semantically associated. Furthermore, as far as pupil size varies in accordance to changes in the imposed level of workload needed to perform a task [31], we expected to reveal physiological differences (i.e. modifications of pupil size) linked to the various kinds of stimuli that were presented in our experimental paradigm (see below for a detailed description).

1.2   The Present Study

The present study aims to disclose feasible differences in pupil diameter due to manipulation of the semantic association between words. We hypothesised that the behaviour of pupil diameter should differ (i.e. the maximal pupil diameter as well as the latency to peak of dilation) in trials in which there is a semantic association between the presented stimuli compared to trials in which there is no association. Moreover, from the perspective of a human-computer symbiosis, monitoring the pupil's behaviours (i.e. through the eye tracker) gives the chance to access and investigate implicit cognitive processing of human beings.

Three pupillary measures were considered in the present study, following the measures of Beatty and Lucero-Wagoner [25]. The first measure was the baseline pupil diameter; namely, the mean pupil diameter that is computed within a pre-measurement temporal interval (i.e. before the presentation of the task stimuli) of constant visual stimulation (i.e. an asterisk placed at the center of the screen is presented during this temporal window). The second measure was the dilation peak, which is defined as the maximal size that the pupil diameter reached in within the time window of interest (i.e. from the task stimuli appearance to the end of the trial). The third measure was the peak latency; that is, the amount of time between the beginning of the time window of interest and the moment in which the pupil reaches its peak of maximal dilation.

We were furthermore interested in studying the effect of covert attention on pupil diameter. Indeed, participants had to maintain their gaze at the center of the screen during the whole experiment. Thus, the changes in size of pupil diameter had to be considered in close relationship to the deployment of covert visual attention. It has been demonstrated in several study how visual stimuli can be processed even if participants are not allowed to look directly at them (i.e. they have to maintain their gaze at the centre of the screen). These studies involved the event-related potentials (i.e. ERPs) technique and paid special attention to a specific ERP component, namely a posterior and controlateral component known as N2pc. The first study in this research series demonstrated that preattentive stimulus information is used to guide attention to task-relevant stimuli in a visual search task [32]. A following study showed dissociations among attention, perception and awareness in visual searches considering the same ERPs component. A relevant visual stimulus can be processed at the level of perception and can be elaborated at a preattentive level (i.e. in an automatic way), but this information can fail to reach awareness [33]. Therefore, pupil diameter could be a measure that mirrors these stages of attentive processing.

2 Method

2.1 Participants

Twenty undergraduate students were involved in this study. The students attended a public university in northern Italy and took part on a voluntary basis. Nine participants were excluded due a high amount of artifacts in the pupil measurements or because their response accuracy to the task was below chance level. We, therefore, considered data from 11 students (5 males) with a mean age of 26 years. All participants had normal or corrected-to-normal vision and gave their informed consent.

2.2   Stimuli

The stimuli were English words selected from the database of the Semantic Priming Project [30]. These 444 words (i.e. 3 words in each trial) were displayed, using E-Prime 2.0 software, in black (i.e. font size 24, style Courier New) on a light gray background (in order to minimise differences in luminance during stimulus presentation) on a 22-inch LCD screen placed 70 cm away from the participants' eyes.

2.3   Procedure

Each trial began with an asterisk at the center of the screen for 1 s and then the asterisk disappeared. A word was presented at the center of the screen for 2 s. This word created a semantic context. The word was then replaced with an asterisk that was displayed alone for 2 s (this time was considered as the temporal window in which the pupil's baseline diameter was computed). Then two words appeared simultaneously. One word was presented to the left and one to the right of the asterisk at 4.6 degrees of visual angle from the center of the screen. The words were displayed up to the moment when a response was provided. Again, an asterisk appeared at the center of the screen for 3 s and then the trial ended. In half of the trials, one of two words was semantically associated with the semantic context word (i.e. target-present trials; for example, *LEGS* as semantic context word and then *UNEVEN* and *KNEES*) while, in the other half, none of the words were associated (i.e. target-absent trials; for example, *SUMMER* as semantic context word and then *EVIDENCE* and *FORCE*; see Fig. 1).

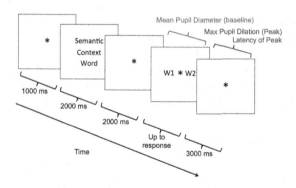

Fig. 1. Graphical depiction of a trial

In each block, three levels of semantic association (i.e. same category, supra-ordinate category, and synonyms) were equiprobable. Each trial was randomly selected. The task was to report the presence of a semantic associated word by pressing the "A" key or its absence by pressing the "L" key (the response keys were counterbalanced across blocks). Accuracy and reaction time were equally stressed. One block of 12 practice trials preceded the two experimental blocks. Before the beginning of each experimental block, a 9-point calibration procedure was performed using a remote eye tracker (RED500, SMI) and Experiment Center software 2.8 (SMI). The pupil diameter was then recorded at a sampling frequency of 500 Hz with iView X 2.8 software (SMI). In each trial, participants were not allowed to look directly at the two eccentric words; but they had to maintain their gaze at the center of the screen. Participants performed two experimental blocks of 72 trials for a total number of 144 trials.

3   Results

We performed a series of mixed-models analyses for the behavioural and physiological data.

3.1   Response Accuracy

The accuracy was analysed by means of generalized linear mixed models (GLMM). The dependent variable was the response accuracy. A series of models was fitted to the data and the fixed effects were: trial type (target present vs. target absent), semantic association type (same category, supraordinate category, synonyms), and their interaction. Subjects was the random effect. The inclusion of this random effect indicates that the variability related to the subjects is taken into account in the models. The model that better fit the data was the model with the two fixed effects (mentioned above) and without the interaction. A main effect of trial type was significant ($b = -0.55$, $z = -3.60$, $p < 0.001$; see Fig. 2). Participants were more accurate when answering to target-absent trials ($M = 86.09$) than in target-present trials ($M = 78.33$). Moreover, there was a significant effect of semantic association type. A multiple-comparison Tukey test was then performed. This test showed a significant difference only between same category and synonym types of semantic association ($b = -0.64$, $z = -3.36$, $p < 0.001$; see Fig. 3). Participants were more accurate in trials with the same category as semantic association type ($M = 86.90$) than in trials with synonyms ($M = 77.77$).

3.2   Response Reaction Time

The response reaction time was analysed by means of linear mixed models. The dependent variable was the response time. A series of models was fit to the data and the fixed effects were: trial type (target present vs. target absent), semantic association type (same category, supraordinate category, synonyms), and their interaction. Subjects was the random effect. The model that better fit the data was the model with the two fixed effects (mentioned above) and without the interaction. A main effect of trial type was significant ($b = -216.23$, $t(1213) = -8.01$, $p < 0.0001$; see Fig. 4). Participants responded faster in target-present trials ($M = 1372.7$ ms) than in target-absent trials ($M = 1591.9$ ms). Moreover,

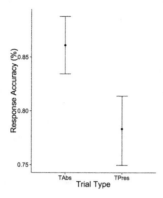

Fig. 2. Response accuracy as a function of trial type

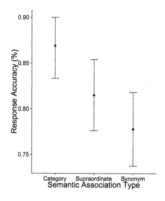

Fig. 3. Response accuracy as a function of semantic association type

there was a significant effect of semantic association type. A multiple-comparison Tukey test was then performed. This test showed a significant difference between same category and synonym types of semantic association only ($b = 144.13$, $z = 4.34$, $p < 0.001$; see Fig. 5). Participants were faster in responding to same category trials ($M = 1402.3\,\text{ms}$) than to synonym trials ($M = 1564.9\,\text{ms}$).

3.3   Physiological Results

We discarded from the analyses all the trials in which participants were looking directly at one of the two words as long as they remained visible on the screen. All the pupil diameter analyses were performed after the pupil signal, which was sampled at $500\,\text{Hz}$, had been pre-processed with a MATLAB (Release 2013a, Mathworks Inc.) script that applied a cubic interpolation and a Butterworth filter (i.e. cut off frequency $= 0.007$; order $= 2$) in order to remove artifacts. Furthermore, a visual inspection of all the graphs, depicting the pupil diameter

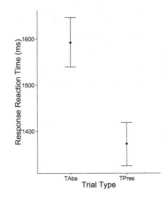

Fig. 4. Response reaction time as a function of trial type

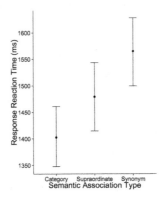

Fig. 5. Response reaction time as a function of semantic association type

for each accepted trial, was performed in order to reject trials in which the maximal pupil dilation was detected in a temporal window of 200 ms after the two-word presentation. In fact, in this temporal window, the peak of dilation could not be ascribed to the pair of words presented, although it could be related to the temporal window of the pupil baseline diameter. By doing so, we preserved the original quality of the signal without down-sampling it. The baseline pupil diameter was calculated in a temporal window of 2 s before the appearance of the word pair. The peak of pupil dilation was identified from the appearance of the task stimuli on the screen to the end of the trial (each trial ended 3 s after the response was provided).

Peak of Maximal Dilation. Peak dilation of the pupil was analysed by means of linear mixed models. The dependent variable was the pupil diameter. A series of models was fit to the data and the fixed effects were: time (mean baseline diameter vs. maximal pupil diameter), trial type (target present vs. target absent),

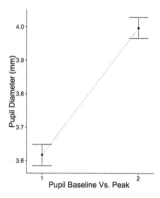

Fig. 6. Pupil size at the baseline and at the peak of dilation

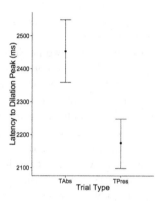

Fig. 7. Temporal window needed by the pupil to reach the peak of dilation

semantic association type (same category, supraordinate category, synonyms), and their interaction. Subjects was considered as the random effect. The model that better fit the data was the model with only time as fixed effect. The main effect of time was significant ($b = 379.60$, $t(2441) = 28.73$, $p < 0.0001$; see Fig. 6). The peak pupil diameter was larger ($M = 3.99\,\text{mm}$) compared to the pupil diameter in the baseline ($M = 3.61\,\text{mm}$).

Peak Latency. Peak latency was analysed by means of linear mixed models. The dependent variable was the time interval needed to reach the dilation peak. A series of models was fit to the data and the fixed effects were: trial type (target present vs. target absent), semantic association type (same category, supraordinate category, synonyms), and their interaction. Subjects was the random effect. The model that better fit the data was the model with only trial type as fixed effect. The main effect of trial type was significant ($b = -275.97$, $t(1215) = -4.785$, $p < 0.0001$). The peak dilation latency was shorter in the target-present trials ($M = 2174.84$ ms) compared to target-absent trials ($M = 2452.85$ ms) (Fig. 7).

4   Discussion

The results showed that our hypothesis regarding possible differences in pupil behaviour was partially confirmed. Indeed, despite the fact that pupil did not show any difference in terms of the maximum diameter size, there was a clear difference in terms of the latency to peak of dilation. In respect to the maximal pupil diameter, cognitive workload typically increases when participants have to perform more difficult tasks [1,7,11,26]. In the current study, participants have to monitor whether there is a semantic association or not without looking directly at the eccentric words. Usually, in order to efficiently process visual stimuli, the cognitive system first needs to deploy the attention over a relevant location of the visual space (i.e. where a target is presented). Furthermore, it is crucial to foveate the relevant information (i.e. look directly at the target stimulus) to achieve deep processing. For this reason in both types of trial, the cognitive workload was higher than it would have been in a free viewing paradigm. Thus, the maximal pupil diameter increases similarly in the temporal window of interest in both type of trials. Moreover, an alternative explanation for this occurrence could be that participants attempted to anticipate whether or not each trial had a semantic association. This is in accordance to what happened to participants that are curious about the answer to trivia questions [5], even though the experimental paradigm was slightly different from the present one. Indeed, in the cited research, authors were checking difference in pupil diameter before and after the presentation of the answer display to monitor the effect of level of curiosity (i.e. high, middle, low) about the answers. But in this research there were not constraints such as to foveate the center of the screen. In the current study, participants are expecting the appearance of the two words in order to provide a response in both kind of trials (i.e. semantic association vs. absence of semantic association). When a task has to be performed regardless of which will be the correct response to provide (i.e. a certain response key for the presence of a semantic association or a different response key for the absence of a semantic association), the cognitive system will be on the alert. The pupil reacts in order to give participants the chance to respond immediately to the presentation of the experimental stimuli.

   In contrast, the latency to the maximal peak of dilation showed a difference. Indeed, the time interval needed to reach the peak of maximal pupil dilation was shorter in trials with a semantic association. Thus, the pupil reflected a diverse cognitive processing in these trials in terms of time needed. The pupil reacted faster (i.e. around 278 ms in a temporal window of around approximately 3000 ms) when the cognitive system detected a semantic association between the presented words. The process of pupil dilation is characterised by a slow trend, so an increment of around 10 % in speed of processing linked to presence of semantic association is considerable. It is likely that participants started, earlier, the pre-attentive stages of processing the stimuli; this is reflected in a quicker global elaboration. This finding supports the idea of a possible dissociation between perception and awareness [33]. Indeed, in about 22 % of trials in which the target was present, participants provided a wrong answer. This means that, even if participants' attentive system had initially processed the presence of a semantic association (i.e. the pupil reacted faster) the elaboration reached the threshold of consciousness in only 78 % of the trials. Indeed, just in two-thirds of these trials, participants provided the correct answer. Higher-level cognitive processing is necessary to accurately perform a task. These results provide new evidence of dissociation among perception (i.e. the brain has processed at perceptive level a target stimulus), attention (i.e. the cognitive system has detected the presence of a target stimulus), and awareness (i.e. the participants are aware of the target stimulus and they are able to report it correctly) as shown in the literature [33].

This last result is really interesting from the perspective of fostering human and machine symbiosis. A system designed to detect changes in pupil diameter, of which humans are not aware, can provide them with information regarding the preconscious processing that is reflected by these variations in size. Thus, human beings could be informed about their preconscious processing through some signals provided by the machine. In this way, humans could be advised, for instance of important visual stimuli (e.g. by highlighting the specific visual area related to the preconscious elaboration that was reflected by the variations in pupil diameter), the representations of which were elaborated at an initial stage but had not reached a sufficient threshold to be completely processed and to become conscious information. The technological applications of these findings could be related, for instance, to information retrieval systems [34, 35] and data searches in complex visual scenarios. Regarding information retrieval, when a user is searching for information, he/she would like to complete the process efficiently and quickly. Thus, the use of pupil diameter information could direct user attention insofar as it has been shown that changes in pupil size are valuable in distinguishing the perceived relevance of Web search results [36]. The system could detect these changes in pupil size and direct (e.g. visually) the user to the more relevant information. Similarly, in order to improve data searches in complex visual scenarios, in which it is known that cognitive workload is high (i.e. some targets among many distracting stimuli), a system could monitor variations in pupil behaviour in order to guide a user in a visual search. Thus, the user could focus just on the most important information and neglect the worthless one. Improved human-machine symbiosis could be achieved when a machine is able to guide observers in order to make their elaboration more efficient. Humans, with this kind of assistance, could concurrently utilize information that is elaborated by their cognitive system at both the preconscious and the conscious level. Obviously the machine has to detect and supply this information online in order to foster humans' cognitive processing as quickly as possible. Furthermore, computers could utilize these changes in pupil behaviour in order to detect cognitive overload. Indeed, high cognitive workload in visual searches is reflected by an increase in pupil diameter. Thus, for example, computers might mitigate these situations by decreasing the amount of information presented to humans. Although some features of the pupil's behaviour suggest its involvement at some level of the semantic processing of words, additional research will be required to disclose the possible role of pupil physiology in the context of a symbiotic relationship between humans and machines.

Acknowledgements. This research was supported by the European Project MIND-SEE, Symbiotic Mind Computer Interaction for Information Seeking (Number: 611570; Call: FP7-ICT-2013-10 Information and Communication Technologies).

References

1. Beatty, J.: Task-evoked pupillary responses, processing load, and the structure of processing resources. Psychol. Bull. **91**(2), 276 (1982)
2. Hess, E.H., Polt, J.M.: Pupil size as related to interest value of visual stimuli. Science **132**(3423), 349–350 (1960)
3. Aboyoun, D.C., Dabbs, J.M.: The Hess pupil dilation findings: sex or novelty? Soc. Behav. Pers. Int. J. **26**(4), 415–419 (1998)
4. Vò, M.L.H., Jacobs, A.M., Kuchinke, L., Hofmann, M., Conrad, M., Schacht, A., Hutzler, F.: The coupling of emotion and cognition in the eye: Introducing the pupil old/new effect. Psychophysiology **45**, 130–140 (2008)
5. Kang, M.J., Hsu, M., Krajbich, I.M., Loewenstein, G., McClure, S.M., Wang, J.T.Y., Camerer, C.F.: The wick in the candle of learning epistemic curiosity activates reward circuitry and enhances memory. Psychol. Sci. **20**(8), 963–973 (2009)
6. Beatty, J., Kahneman, D.: Pupillary changes in two memory tasks. Psychon. Sci. **5**(10), 371–372 (1966)
7. Iqbal, S.T., Zheng, X.S., Bailey, B.P.: Task-evoked pupillary response to mental workload in human-computer interaction. In: CHI'04 Extended Abstracts on Human Factors in Computing Systems, pp. 1477–1480. ACM, April 2004
8. Van Orden, K.F., Jung, T.P., Makeig, S.: Combined eye activity measures accurately estimate changes in sustained visual task performance. Biol. Psychol. **52**(3), 221–240 (2000)
9. Hess, E.H., Polt, J.M.: Pupil size in relation to mental activity during simple problem-solving. Science **143**(3611), 1190–1192 (1964)
10. Ahern, S., Beatty, J.: Pupillary responses during information processing vary with Scholastic Aptitude Test scores. Science **205**, 1289–1292 (1979)
11. Hyönä, J., Tommola, J., Alaja, A.M.: Pupil dilation as a measure of processing load in simultaneous interpretation and other language tasks. Q. J. Exp. Psychol. **48**(3), 598–612 (1995)
12. Just, M.A., Carpenter, P.A.: The intensity dimension of thought: pupillometric indices of sentence processing. Can. J. Exp. Psychol./Rev. canadienne de psychologie expérimentale **47**(2), 310 (1993)
13. Kahneman, D., Beatty, J.: Pupil diameter and load on memory. Science **154**(3756), 1583–1585 (1966)
14. Porter, G., Troscianko, T., Gilchrist, I.D.: Effort during visual search and counting: insights from pupillometry. Q. J. Exp. Psychol. **60**, 211–229 (2007)
15. Van Gerven, P.W., Paas, F., Van Merriënboer, J.J., Schmidt, H.G.: Memory load and the cognitive pupillary response in aging. Psychophysiology **41**(2), 167–174 (2004)

16. Janisse, M.P., Bradley, M.T.: Deception information and the pupillary response. Percept. Mot. Skills **50**(3), 748–750 (1980)
17. Lubow, R.E., Fein, O.: Pupillary size in response to a visual guilty knowledge test: New Technique for detection of deception. J. Exp. Psychol. Appl. **2**(2), 164–177 (1996)
18. Dionisio, D.P., Granholm, E., Hillix, W.A., Perrine, W.F.: Differentiation of deception using pupillary responses as an index of cognitive processing. Psychophysiology **38**(2), 205–211 (2001)
19. Jacucci, G., Spagnolli, A., Freeman, J., Gamberini, L.: Symbiotic interaction: a critical definition and comparison to other human-computer paradigms. In: Jacucci, G., Gamberini, L., Freeman, J., Spagnolli, A. (eds.) Symbiotic 2014. LNCS, vol. 8820, pp. 3–20. Springer, Heidelberg (2014)
20. Licklider, J.C.R.: Man-computer symbiosis. Hum. Factors Electron. IRE Trans. **1**, 4–11 (1960)
21. Spagnolli, A., Guardigli, E., Orso, V., Varotto, A., Gamberini, L.: Measuring user acceptance of wearable symbiotic devices: validation study across application scenarios. In: Jacucci, G., Gamberini, L., Freeman, J., Spagnolli, A. (eds.) Symbiotic 2014. LNCS, vol. 8820, pp. 87–98. Springer, Heidelberg (2014)
22. Negri, P, Gamberini, L, Cutini, S.: A Review of the Research on Subliminal Techniques for Implicit Interaction in Symbiotic Systems. In: Jacucci, G., Gamberini, L., Freeman, J., Spagnolli, A. (eds.) Symbiotic 2014. LNCS, vol. 8820, pp. 47–60. Springer, Heidelberg (2014)
23. Barral, O., Aranyi, G., Kouider, S., Lindsay, A., Prins, H., Ahmed, I., Jacucci, G., Negri, P., Gamberini, L., Pizzi, D., Cavazza, M.: Covert persuasive technologies: bringing subliminal cues to human-computer interaction. In: Persuasive Technology, pp. 1–12. Springer International Publishing (2014)
24. Aranyi, G., Kouider, S., Lindsay, A., Prins, H., Ahmed, I., Jacucci, G., Negri, P., Gamberini, L., Pizzi, D., Cavazza, M.: Subliminal cueing of selection behavior in a virtual environment. Presence: Teleoperators Virtual Environ. **23**(1), 33–50 (2014)
25. Beatty, J., Lucero-Wagoner, B.: The pupillary system. Handb. Psychophysiol. **2**, 142–162 (2000)
26. Holmqvist, K., Nyström, M., Andersson, R., Dewhurst, R., Jarodzka, H., Van de Weijer, J.: Eye Tracking: A Comprehensive Guide to Methods and Measures. Oxford University Press, Oxford (2011)
27. Laeng, B., Sirois, S., Gredebäck, G.: Pupillometry a window to the preconscious? Perspect. Psychol. Sci. **7**(1), 18–27 (2012)
28. Whipple, B., Ogden, G., Komisaruk, B.R.: Physiological correlates of imagery-induced orgasm in women. Arch. Sex. Behav. **21**(2), 121–133 (1992)
29. Loewenfeld, I.: The Pupil: Anatomy, Physiology, and Clinical Applications. Detroit MI Wayne State University Press, Detroit (1993)
30. Hutchison, K.A., Balota, D.A., Neely, J.H., Cortese, M.J., Cohen-Shikora, E.R., Tse, C.S., Yap, M.J., Bengson, J.J., Niemeyer, D., Buchanan, E.: The semantic priming project. Behav. Res. Methods **45**(4), 1099–1114 (2013)
31. Kahneman, D.: Attention and Effort. Prentice-Hall, Englewood Cliffs (1973)
32. Luck, S.J., Hillyard, S.A.: Electrophysiological correlates of feature analysis during visual search. Psychophysiology **31**(3), 291–308 (1994)
33. Woodman, G.F., Luck, S.J.: Dissociations among attention, perception, and awareness during object-substitution masking. Psychol. Sci. **14**(6), 605–611 (2003)
34. Manning, C.D., Raghavan, P., Schütze, H.: Introduction to Information Retrieval, pp. 1–6. Cambridge University Press, Cambridge (2008)

35. Athukorala, K., Hoggan, E., Lehti, A., Ruotsalo, T., Jacucci, G.: Information-seeking behaviors of computer scientists: challenges for electronic literature search tools. Proc. Am. Soc. Inf. Sci. Technol. **50**(1), 1–11 (2013)
36. Oliveira, F.T., Aula, A., Russell, D.M.: Discriminating the relevance of web search results with measures of pupil size. In: Proceedings of the SIGCHI Conference on Human Factors in Computing Systems, pp. 2209–2212. ACM (2009)

Demos and Posters

Querying and Display of Information: Symbiosis in Exploratory Search Interaction Scenarios

Barış Serim[✉]

Helsinki Institute for Information Technology HIIT,
Department of Computer Science,
University of Helsinki, Helsinki, Finland
baris.serim@hiit.fi

Abstract. This paper examines potential interaction aspects related to querying and the display of information in exploratory search scenarios with a particular focus on user state and interactive visualization. Exploratory search refers to a specific type of information seeking that is open-ended, continuous and evolving. The evolving nature of exploratory search also provides the computer with sequential data that can be used to estimate user state and intention as the search unfolds. In this setting, the system supports querying by relying on user's *pointing* actions, *sequential organization* of user interaction and *query metadata*. The system also adapts the display of information by determining the *timing* and *visual representation*. The paper illustrates potential interactions that employ new input modalities such as eye gaze and physiological signals. The paper concludes by discussing the possible functions of interactive visualization regarding querying and the display of information.

Keywords: Interaction design · Physiological input · Gaze input · Interactive visualization · Search context · Exploratory search

1 Introduction

Digital/web search has become the primary tool for most of the information seeking tasks. Most search activity can be classified as basic look-up, searches that are known-item or fact retrieval tasks. Yet, long term search activities, such as exploring new domains of knowledge, are often more complex. Information seeking in such long term exploratory search often involves evaluation, comparison and synthesis of results and iterative querying. However, as expressed by Marchionini such open ended information needs are not well addressed in today's search engines that are oriented towards high precision rather than maximizing the number of possibly relevant objects [1]. Related to this observation is the questioning of the dominant interaction model for search tools, "query-response paradigm", and the reevaluation of guidelines for the design of systems to support exploratory search [2].

© Springer International Publishing Switzerland 2014
G. Jacucci et al. (Eds.): Symbiotic 2014, LNCS 8820, pp. 115–120, 2014.
DOI: 10.1007/978-3-319-13500-7_9

Typical query-response pair illustrates some common interactive qualities of search interfaces. Consider this sequence of events that occur during a basic look-up task:

A user types a query to a predefined search box in the graphical user interface and submits her query. The system responds by returning a set of results that are displayed as a list, sorted by their relevance as computed by the search engine.

The sequence reveals many aspects of interaction that include but are not limited to (a) the input modality (typing), (b) the initiation mechanism (explicitly initiated by user), and (c) the timing (concurrent), (d) presentation (textual) and (e) layout (list) of search results. In HCI various prototypes and design frameworks emerged that took a different approach to one or more of the various aspects listed above. This paper aims to identify future interactions with a particular focus on information visualization and various input modalities (such as EEG, EDR, fEMG, eye gaze and pupillometry which are currently positioned as peripheral in respect to more established and precise input techniques). Academic search, in which open ended and evolving searches is common, is used as a case.

Input modalities such as gaze and other physiological signals expand the resources available to the system for inferring the state of user [3], increasing the capability of the system for collaboration [4]. In search scenarios such input allows an increased ability of the computer in assigning user responses to information items. The increased awareness of user response also implies possible changes regarding how the user performs queries and how the system presents information in future search interfaces.

2   Querying

A potential outcome of human computer symbiosis in exploratory search scenarios is the possibility of performing search without having to articulate precise queries. In this setting, the system carries out the necessary work of formulating the query by possibly relying on user's spatial references (pointing), the sequential organization of user actions and query metadata.

Pointing. Pointing, as a spatial referencing technique, allows directing attention to objects by using contextual information and is widely used in graphical interfaces through cursor and touch input. Pointing is also a possible method for facilitating the easy construction of queries as opposed to using typed queries. In most search interfaces pointing is used for selecting specific items, filters or links. Apart from these, several visual interfaces allow pointing to objects such as keywords or document surrogates for open ended querying. IntentRadar [5] features radially organized keywords as a representation of the user query and enables dragging the keywords for directing the search. Apolo [6] enables pointing to surrogates of scientific publications for query, with the system populating the visual interface with similar publications in response. In addition to pointing single items, pointing to multiple items or regions is possible by using interactive visualization and query relaxation techniques [7]. In common, these examples allow the user to form queries by using items that are readily available in the interface and assign the task of formulating the query to the system. On the other hand, possible forms of pointing are not limited to mouse or touch input. Spatial references can take many forms such as different hand postures, gaze and orientation of head, body and arms. Future interfaces can build on these different forms of indications for diversifying the possible modes of explicit inputs and for tracking user's attention.

Sequential organization. Interaction during information seeking involves consecutive user actions while interacting with the user interface. The sequence of these actions constitutes part of what is called the search context and have been used to infer user preferences and to expand or disambiguate queries. Traditionally, most implicit mining of user data involved actions that are common to graphical user interfaces such as scrolling, selecting, navigating, saving, deleting [8], with few examples using alternative inputs such as gaze [9]. Incorporation of peripheral physiological input, allows more detailed information related to user state and interest level during the interaction, without having to rely on user's explicit feedback. At the same time, the sequential organization does not always imply relation. Thus, one possible research area would be to identify gaps and adjacency pairs during interaction, or to provide the user the necessary means to designate search sequences.

Query metadata. In addition to sequential organization, the queries can be contextualized by using inputs that are concurrent to the primary input (Fig. 1). These can include both the metadata of the primary input (e.g. the pace of typing when entering a query), or peripheral input that accompany the primary input (e.g. accompanying eye gaze or physiological signals while the user is typing a query). These metadata can be used to clarify user's query, similar to the role of intonation, pitch and amplitude for interpreting participants' utterances in daily conversation [10]. An HCI example of using input metadata is registering pressure information while typing on a small keyboard to explicitly control the level of input uncertainty [11].

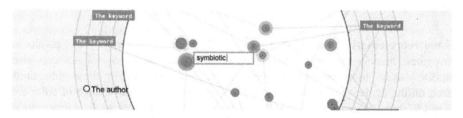

Fig. 1. Querying actions like typing can be contextualized by using immediate or concurrent input. The figure shows a possible interaction scenario, in which the query box follows user's gaze input. In this case, the space and the nearby elements constitute part of the context of the typed query.

3 Display of Information

Another outcome of human computer symbiosis is adapting the display of information to the user state by relying on detailed information about user's attention level and response. The system utilizes this information to determine the timing and visual representation of information items.

Timing. In typical search interfaces the display of the results is concurrent: all the results are displayed at once. In contrast, future systems can use time order expressively to highlight and relate certain items within the retrieved result set by the order of their introduction. Gaze input in these cases can be used to detect user's attention levels, in turn affecting the pace of display.

Besides using timing in response to a query, a radical alternative is to eliminate queries and proactively retrieve information based on user's changing context. In HCI various design frameworks exist such as non-command [12], mixed-initiative [13] and attentive user interfaces [14] that promote information retrieval without users having to explicitly query, but rather rely on software agents [15]. Such proactive display of information, however, potentially causes the problem of distracting the user from the task at hand. As Allen [16] has noted, as opposed to fixed initiative systems in which the initiation of interaction is well-defined, in mixed initiative systems the agents should decide on the appropriate time of starting the interaction. User actions, as well as gaze and other physiological signals in this case can potentially provide necessary triggers for the timing of information display.

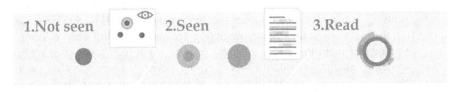

Fig. 2. The document surrogates can be visualized to indicate different attention levels and responses a document goes through. The figure shows illustrative visualizations of a document surrogate which (1) is not yet seen, (2) is seen and (3) has been opened and the document it represents has been read. Different color saturations can show different levels of attention and arousal.

Visual representation. Search interfaces usually represent retrieved set of results in surrogates, namely the representation of original documents. Surrogates can also indicate various user actions performed by the user. Traditionally these indications relied on the explicit input of the user. Typical examples are the change of color for visited Web pages in a search result list and the display of actions such as forwarding, replying and reading for emails. EEG and other physiological inputs provide the opportunity to visually distinguish information items without any explicit input (Fig. 2), by tracking user attention (whether the surrogate is gazed at by the user) and physiological response (how the interaction with the document affected the cognitive load and arousal level of the user). Interaction history can also enable tailoring the layout of surrogates in the visual space by taking the user interaction history into account. A possible example is changing the position of a document in a graph after user interaction, by registering what the user specifically engaged within the document. In this scenario possible clues for the change include sequential organization of search as well as various physiological inputs recorded while the user is engaged with the different parts of the document.

4   Discussion: Roles of Visualization

This section identifies possible roles for the visualization of items during search regarding the querying and display of information examined above.

As a resource for pointing. Visual display of information provides set of items that the user can point to for explicit querying. In addition to querying, visualization enables contextualizing the search through sequential organization or query metadata by pointing to the items prior to or during the query. These pointing actions can be used to indicate not only the items themselves but the specific user interaction history with the item (e.g. what the user has specifically found relevant within the item). In this case the registration of user's previous response allows referring to the past context of interaction.

Orient the user. Search results from various stages of a search session can be visualized to orient the user by juxtaposing newly retrieved information items with familiar ones. Visual representation of the previously interacted items, in this case, indicates how familiar the user is with an item and her past implicit response during interaction.

Prioritize events and information items. Visualization enables prioritizing information items and events by displaying them in different visual areas and with different visual features. Together with timing, visualization enables notifying the user of possible actions that are initiated by the system such as adding, removing or highlighting items.

Represent system estimation of user intention. The systems estimation of user state or intention can be conveyed through visual representation, for the user to make sense of system actions and repair any miscommunications.

5   Conclusion and Future Work

The paper identified potential design directions regarding querying and display of information in future search scenarios by focusing on new input modalities and information visualization. These design directions also described how human computer symbiosis can be furthered with the increasing role of computers for formulating queries and adapting the display to the user. The review of the above interactive aspects is by no means exhaustive but is intended to spark a discussion for future design. Future work will focus on prototyping these different dimensions for realistic search scenarios.

Acknowledgements. This work has been partly supported by MindSee (FP7 – ICT; Grant Agreement # 611570).

References

1. Marchionini, G.: Exploratory search: from finding to understanding. Commun. ACM **49**(4), 41–46 (2006)
2. White, R.W., Roth, R.A.: Exploratory search: beyond the query-response paradigm. In: Synthesis Lectures on Information Concepts, Retrieval, and Services. Morgan & Claypool (2009)

3. Barral O., Jacucci G.: Applying physiological computing methods to study psychological, affective and motivational relevance. In: Proceedings of Symbiotic'14. Springer Berlin Heidelberg (2014)

4. Jacucci, G., Spagnolli, A., Freeman, J., Gamberini, L.: Symbiotic interaction: a critical definition and comparison to other human-computer paradigms. In: Proceedings of Symbiotic'14. Springer Berlin Heidelberg (2014)

5. Ruotsalo, T., Peltonen, J., Eugster, M., Głowacka, D., Konyushkova, K., Athukorala, K., Jacucci, G., Kaski, S.: Directing exploratory search with interactive intent modeling. In: Proceedings of CIKM'13, 1759-1764. ACM Press, New York (2013)

6. Chau, D.H., Kittur, A., Hong, J.I., Faloutsos, C.: Apolo: making sense of large network data by combining rich user interaction and machine learning. In: Proceedings of CHI'13, pp. 167–176. ACM Press, New York (2013)

7. Heer, J., Agrawala, M., Willett, W.: Generalized selection via interactive query relaxation. In: Proceedings of CHI'08, pp. 959–968. ACM Press, New York (2008)

8. Kelly, D., Teevan, J.: Implicit feedback for inferring user preference: a bibliography. In: ACM SIGIR Forum, vol. 37, issue 2, pp. 18–28. ACM Press, New York (2003)

9. Maglio, P.P., Barrett, R., Campbell, C.S., Selker, T.: SUITOR: An attentive information system. In: Proceedings of IUI'00, pp. 169–176. ACM Press, New York (2000)

10. Tannen, D.: Conversational style. Ablex, Norwood (1984)

11. Weir, D., Pohl, H., Rogers, S., Vertanen, K., Kristensson, P.O.: Uncertain text entry on mobile devices. In: Proceedings of CHI'14, pp. 2307–2316. ACM Press, New York (2014)

12. Nielsen, J.: Noncommand user interfaces. Commun. ACM **36**(4), 83–99 (1993). ACM Press, New York

13. Horvitz, E.: Principles of mixed-initiative user interfaces. In: Proceedings of CHI'99, pp. 159–166. ACM Press, New York (1999)

14. McCrickard, D.S., Chewar, C.M.: Attuning notification design to user goals and attention costs. Commun. ACM **46**(3), 67–72 (2003). ACM Press, New York

15. Rhodes, B.J., Maes, P.: Just-in-time information retrieval agents. IBM Syst. J. **39**(4), 685–704 (2000)

16. Allen, J.E., Guinn, C.I., Horvitz, E.: Mixed-initiative interaction. IEEE Intell. Syst. Appl. **14**(5), 14–23 (1999)

A Reinforcement Learning Approach to Query-Less Image Retrieval

Sayantan Hore, Lasse Tyrvainen, Joel Pyykko, and Dorota Glowacka[✉]

Helsinki Institute for Information Technology HIIT,
Department of Computer Science, University of Helsinki, Helsinki, Finland
glowacka@cs.helsinki.fi

Abstract. Search algorithms in image retrieval tend to focus exclusively on giving the user more and more similar images based on queries that the user has to explicitly formulate. Implicitly, such systems limit the users exploration of the image space and thus remove the potential for serendipity. Thus, in recent years there has been an increased interest in developing exploration–exploitation algorithms for image search. We present an interactive image retrieval system that combines Reinforcement Learning together with a user interface designed to allow users to actively engage in directing the search. Reinforcement Learning is used to model the user interests by allowing the system to trade off between exploration (unseen types of image) and exploitation (images the system thinks are relevant).

1 Introduction

In recent years, image retrieval techniques operating on meta-data, such as textual annotations or user-specified tags, have become the industry standard for retrieval from large image collections, e.g. Google Image Search. This approach works well with sufficiently high-quality meta-data, however, with the explosive growth of image collections, evidenced by Facebook users averaging 350 million photo uploads per day during the fourth quarter of 2012 and Instagram reporting an average of 55 million, it has became apparent that tagging new images quickly and efficiently is not always possible [12]. Secondly, even if instantaneous high-quality image tagging was possible, there are still many instances where image search by query is problematic. It might be easy for a user to define their query if they are looking for an image of a cat but how do they specify that the cat should be of a very particular shade of ginger with sad looking eyes. A solution to a situation where the user is unable to specify the required content through tags or other image properties is content-based image retrieval (CBIR).

The first CBIR experiments date back to 1992 [8] and since then many CBIR image retrieval systems have been developed [4,10,15,17]. Early experiments show that the process of image retrieval can be greatly improved by employing relevance feedback [20] through actively involving the user into the search loop and utilizing his knowledge in the iterative search process [3,11,18]. However, evidence from user studies indicates that relevance feedback leads to a context

© Springer International Publishing Switzerland 2014
G. Jacucci et al. (Eds.): Symbiotic 2014, LNCS 8820, pp. 121–126, 2014.
DOI: 10.1007/978-3-319-13500-7_10

trap, i.e. after a few iterations of feedback users have specified their context so strictly that the system is unable to propose anything new while the users are trapped within the present set of results and can only exploit a limited area of information space [9]. Faceted search [19] was an attempt to solve the problem of context trap by using global features and allowing the user to explore a collection of images by applying multiple filters. However, the number of global features can be very large thus forcing the user to select from a large amount of options, which can make the whole process inconvenient and cognitively demanding.

Employing various *exploration/exploitation* strategies into relevance feedback has been another attempt at avoiding the context trap. The exploitation step aims at returning to the user the maximum number of relevant images in a local region of the feature space, while the exploration step aims at driving the search towards different areas of the feature space in order to discover not only relevant images but also informative ones. This type of systems control dynamically, at each iteration, the selection of displayed images. For example, in [16] a mass-zoom algorithm is proposed to modulate how much the displayed set of images should be concentrated on the images assessed as the most relevant by the user. In [13], a method is considered that exploits the information obtained from the relevance scores as computed by a nearest neighbour approach in the exploitation step. Similarly, reinforcement learning is a promising approach that can allow the system to utilise the search context in relevance feedback and, at the same time, avoid the context trap by trading between exploitation and exploration. Reinforcement learning has been applied in various areas of information retrieval and recommender systems, including image retrieval [2,6]. However, most of the image retrieval systems employing exploration - exploitation strategies [13,16,19] also require the user to, at least in the first iteration, formulate some kind of query to initiate the search. In this paper, we describe the design of a query-less image search system incorporating state of the art reinforcement learning techniques to allow the system to efficiently balance between exploration and exploitation.

The primary goal of the system is to assist the user in finding images in a database of unannotated images quickly and effectively without typing any queries. Reinforcement learning (RL) methods as well as interactive interface allow the user to assign relevance scores to the displayed images by moving a pointer on a sliding bar. Through assigning scores to the presented images, the user can direct the search according to their interest, while the inbuilt RL mechanism helps the system to form a model of user's interests and suggest appropriate images in the next search iteration. The user modeling process is restarted for each new search session to avoid the issue of "over-personalization".

2   Interface Design

The main idea behind the interactive interface is that instead of typing queries related to the desired image, the user navigates through the contents by indicating how "close" or "similar" the displayed images are to their ideal target image.

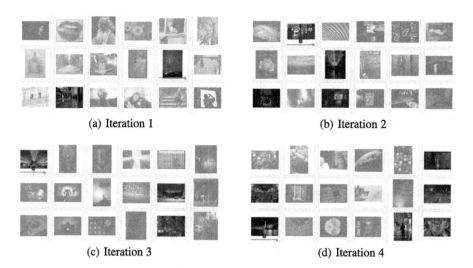

(a) Iteration 1

(b) Iteration 2

(c) Iteration 3

(d) Iteration 4

Fig. 1. The first four iterations of an example search for "City by night". In the first iteration the user marks two images as relevant. As the search progresses more and more images related to the user's search intent are displayed.

At the next iteration, the user is presented with a new set of images. The newly displayed collage contains more images with features that are of interest to the user. The user interface and an example search session are presented in Fig. 1.

The search starts with a display of a collage of images that are sampled uniformly at random from the database. In order to ensure that the initial set of images is a good representation of the entire image space, we cluster all the images in k clusters, where k is the number of displayed images and then we sample an image from each cluster. Our pre user study experiments show that this technique provides a good starting point for the search. When the mouse hovers over an image, a slide bar appears at the bottom allowing the user to rate the relevance of that particular image. The feedback ranges from 0 to 1, where 0 indicates that a given image is very different from the ideal target and 1 indicates that a given image is very similar to the ideal target. The user can rate as many of the displayed images as they like. Images that were not rated by the user are assumed to have score of 0. After each iteration, new images are displayed. Each image can be displayed to the user at most once throughout the entire search session, i.e. an image displayed at a given iteration cannot be displayed at subsequent ones. The search continues until the user is satisfied with the results.

We illustrate the interface and interaction design through a walkthrough that exemplifies a real image searching task. In our example, the user wants to find an image to illustrate an article about "city by night". At the start of the search (Fig. 1a), the user is presented with a collage of images uniformly selected from the image database and he marks the fifth image in the second row and the second image in the third row as being close to their ideal target image. The user moves to the next iteration by pressing the "Next" button at the top of the page. In the second iteration (Fig. 1b), more images related to "evening" or "night" are presented. Again, the user selects four images that are close to what he is looking for. In iterations 3 and 4 (Fig. 1c and d), more and more relevant images are presented to the user and the user can narrow down his search by selecting more specific images.

3   User Modelling

Throughout the search session, the system builds the user model and based on it selects k images to present to the user at each iteration. We assume that the user is looking for an image or a set of images related to their interest. The user feedback is given by a relevance score $y \in [0, 1]$, where 1 indicates that the image is of high relevance. The formal iteration protocol is as follows:

- The system selects k images and presents them to the user.
- The user provides relevance scores $y_i \in [0, 1], i = 1, \ldots, k$ for the displayed images. Images that have not been rated by the user are assumed to have relevance score of 0.

In order to help the user to explore the image space, we use Gaussian Process (GP) bandits [14], an algorithm that has already been proven to work well in information retrieval applications [7]. GP bandits are an extension of simple bandit algorithms [1] to a case where there are dependencies across arms, which in our system translates into handling similarities between images based on their features. At each iteration, the algorithm suggests new images based on the user feedback from previous iterations. The algorithm uses function f that makes predictions with regards to the relevance of all the images in the database to the user's interests. When selecting the next set of images to display, the system might simply select the images with the highest estimated relevance score. But since the estimate of f may be inaccurate, this exploitative choice might be suboptimal. Alternatively, the system might exploratively select an image for which the user feedback improves the accuracy of f, enabling better image selections in subsequent iterations.

Let us assume that $\mathscr{X}$ is a space whose elements are images $x_1, \ldots, x_n$. Images seen up to time t are $x_1, \ldots, x_{t-1}$ with relevance scores $y_1, \ldots, y_{t-1}$. The GP posterior at time t after seeing data $(x_1, y_1), \ldots, (x_{t-1}, y_{t-1})$ has mean $\mu_t(x)$:

$$\mu_t(x) = k_{t-1}(x)^T C_{t-1}^{-1} y_{t-1} \qquad (1)$$

with variance $\sigma_t^2(x)$:

$$\sigma_t^2(x) = \kappa(x, x) - k_{t-1}(x)^T C_{t-1}^{-1} k_{t-1}(x) \qquad (2)$$

where κ denotes a kernel defined on pairs of elements of $\mathscr{X}$: $\kappa = C(x, x)$ and C_{t-1} is the covariance matrix between the elements of $\mathscr{X}$ seen up to time t, while $k_t = (C(x_1, x_1), \ldots, C(x_{t-1}, x_{t-1}))$. We used the Gaussian kernel with the Hellinger distance between images:

$$\kappa(x_i, x_j) = \exp\left(-\frac{\|\, x_i - x_j\,\|}{2l^2}\right) \tag{3}$$

We set the value of l to 0.5, which was selected during the pre-user study experiments.

The algorithm aims to optimally balance exploration and exploitation by selecting images to display according to the following formula:

$$x_t = argmax_{x \in \mathcal{X}}\{f_{t-1}(x) = \mu_{t-1}(x) + B(t)\sigma_{t-1}(x)\} \tag{4}$$

This can be seen as active learning where we want to learn accurately in regions where f looks good, while ignoring other areas. The $B(t)$ term balances exploration and exploitation: the bigger it gets, the more it favours points with high $\sigma_t(x)$. The value of $B(t)$ was set to $B(t) = 0.002$ was selected during the pre-user study experiments as it was the optimal one. The objective of the algorithm is to find as quickly as possible a good approximation $f^\star$ of the maximum of f. At each iteration, we learn new information which enables us to improve our approximation of $f^\star$ over time. Initial experimental results involving a set of user studies can be found in [5].

4   Summary and Conclusion

To support users in directing the exploration of the image space during search, we developed an interactive CBIR system that applies exploration-exploitation trade-off with user feedback. Instead of typing queries, users can explore the image space by scoring images currently on display. Such interactions result in predictions of new images matching the user interest at the current search iteration. The system does not require a dataset with tagged images, the users do not need to think of elaborate queries when searching for an image, it supports users' exploratory behaviour when dealing with an unknown dataset of images or loosely defined search target.

Acknowledgements. The project was supported by The Finnish Funding Agency for Innovation (under projects Re:Know and D2I) and by the Academy of Finland (under the Finnish Centre of Excellence in Computational Inference).

References

1. Auer, P., Cesa-Bianchi, N., Fischer, P.: Finite-time analysis of the multiarmed bandit problem. Mach. Learn. **47**, 235–256 (2002)
2. Auer, P., Hussain, Z., Kaski, S., Klami, A., Kujala, J., Laaksonen, J., Leung, A.P., Pasupa, K., Shawe-Taylor, J.: Pinview: implicit feedback in content-based image retrieval. JMLR Workshop Conf. Proc. **11**, 51–57 (2010)

3. Cox, I., Miller, M., Minka, T., Papathomas, T., Yianilos, P.: The Bayesian image retrieval system, PicHunter: theory, implementation, and psychophysical experiments. Image Process. **9**(1), 20–37 (2000)
4. Datta, R., Li, J., Wang, J.: Content-based image retrieval: approaches and trends of the new age. In: Multimedia Information Retrieval, pp. 253–262. ACM (2005)
5. Głowacka, D., Hore, S.: Balancing exploration-exploitation in image retrieval. In: Proceedings of UMAP 2014 Posters, Demonstrations and Late-Breaking Results (2014)
6. Głowacka, D., Shawe-Taylor, J.: Content-based image retrieval with multinomial relevance feedback. In: Proceedings of ACML, pp. 111–125 (2010)
7. Guiver, J., Snelson, E.: Learning to rank with softrank and Gaussian processes. In: Proceedings of SIGIR, pp. 259–266 (2008)
8. Kato, T., Kurita, T., Otsu, N., Hirata, K.: A sketch retrieval method for full color image database-query by visual example. In: Pattern Recognition. Computer Vision and Applications, pp. 530–533 (1992)
9. Kelly, D., Fu, X.: Elicitation of term relevance feedback: an investigation of term source and context. In: Proceedings of SIGIR (2006)
10. Kosch, H., Maier, P.: Content-based image retrieval systems-reviewing and benchmarking. JDIM **8**(1), 54–64 (2010)
11. Laaksonen, J., Koskela, M., Laakso, S., Oja, E.: Picsom-content-based image retrieval with self-organizing maps. Pattern Recogn. Lett. **21**(13), 1199–1207 (2000)
12. Pham, T.-T., Maillot, N.E., Lim, J.-H., Chevallet, J.-P.: Latent semantic fusion model for image retrieval and annotation. In: Proceedings of CIKM (2007)
13. Piras, L., Giacinto, G., Paredes, R.: Enhancing image retrieval by an exploration-exploitation approach. In: Perner, P. (ed.) MLDM 2012. LNCS (LNAI), vol. 7376, pp. 355–365. Springer, Heidelberg (2012)
14. Rasmussen, C.E., Williams, C.K.I.: Gaussian Processes for Machine Learning. MIT Press, Cambridge (2006)
15. Smeulders, A., Worring, M., Santini, S., Gupta, A., Jain, R.: Content-based image retrieval at the end of the early years. Pattern Anal. Mach. Intell. **22**(12), 1349–1380 (2000)
16. Suditu, N., Fleuret, F.: Iterative relevance feedback with adaptive exploration/exploitation trade-off. In: Proceedings of CIKM (2012)
17. Veltkamp, R.C., Tanase, M.: Content-based image retrieval systems: a survey. Department of Computing Science, Utrecht University (2002)
18. Villegas, M., Leiva, L.A., Paredes, R.: Interactive image retrieval based on relevance feedback. In: Sappa, A.D., Vitrià, J., Multimodal Interaction in Image and Video Application, vol. 48, pp. 83–109. Springer, Heidelberg (2013)
19. Yee, K.-P., Swearingen, K., Li, K., Hearst, M.: Faceted metadata for image search and browsing. In: Proceedings of CHI, pp. 401–408 (2003)
20. Zhou, X., Huang, T.: Relevance feedback in image retrieval: a comprehensive review. Multimed. Syst. **8**(6), 536–544 (2003)

A Multi-touch Interface for Multi-robot Path Planning and Control

Salvatore Andolina[1]([⊠]) and Jodi Forlizzi[2]

[1] HIIT and Department of Computer Science,
University of Helsinki, Helsinki, Finland
`salvatore.andolina@cs.helsinki.fi`
[2] HCII and School of Design, Carnegie Mellon University, Pittsburgh, USA
`forlizzi@cs.cmu.edu`

Abstract. In the last few years, research in human-robot interaction has moved beyond the issues concerning the design of the interaction between a person and a single robot. Today many researchers have shifted their focus toward the problem of how humans can control a multi-robot team. The rising of multi-touch devices provides a new range of opportunities in this sense. Our research seeks to discover new insights and guidelines for the design of multi-touch interfaces for the control of biologically inspired multi-robot teams. We have developed an iPad touch interface that lets users exert partial control over a set of autonomous robots. The interface also serves as an experimental platform to study how human operators design multi-robot motion in a pursuit-evasion setting.

Keywords: Human-robot interaction · Multi-agent path planning · Biologically inspired motion · Multi-touch interface

1 Introduction

The field of human-robot interaction has evolved beyond issues concerning the design and development of one person controlling one robot. Today, robots and humans routinely collaborate in multi-robot teams. Robotics has advanced such that teams of robots can be fully autonomous, completing multiple tasks in parallel. In addition to interaction with a group of agents that is fully autonomous, a group of agents could be fully controlled by the user, or some combination that exists in between.

In this paper, we present an experimental platform that helps to understand how motion and path planning for multi-robot swarms should be designed. We are especially interested in cases where a human exerts partial control over multiple robots. In [1] researchers explore biologically inspired motion by using an approach where the input is initiated by an operator, applies to the motion of a subset of robots, and in turn, affects the motion of other robots. In our work, we developed a multi-touch interface to study how decentralized control over a set of biologically inspired agents can be used to reach high-level pursuit-evasion strategies. In this paper, we describe the design and development of the interface.

© Springer International Publishing Switzerland 2014
G. Jacucci et al. (Eds.): Symbiotic 2014, LNCS 8820, pp. 127–132, 2014.
DOI: 10.1007/978-3-319-13500-7_11

2   Background

Research on path-planning has explored a number of algorithms for fully autonomous systems. Some of them, such as discrete search methods [5,12,13], work for path planning, but do not scale well to multiple robots. Other optimized methods [7], while working for multiple robots, do not function well in complex environments.

Multi-robot motion design research has often leveraged pursuit-evasion tasks. Different algorithms have been designed and tested in simulation to plan the path of pursuers and evaders, ranging from two pursuers are searching for one evader [16] to multiple pursuers searching for multiple evaders [3,18].

Biological motion has also been a source of inspiration for those researching multi-robot path planning. The study of the collective behavior found in nature has led to attempts to create computational models to characterize and replicate that behavior. Couzin et al. [4] offers a popular bio-inspired model that describes three circular zones around an agent: repulsion, orientation, and attraction. The motion of each robot is calculated based on the presence or absence of other individuals on each of these zones. This globally generates a collective behavior that resembles the behavior of a real swarms.

A number of interfaces and interaction studies can be found exploring how to control a single and multiple robots [2,6,8,10,11]. A special case is that of multi-touch interfaces. While surface and direct touch computing can offer challenges in the forms of less traditional input than the keyboard and mouse, it offers a direct means of controlling elements through multitouch freehand gestures [17]. A number of systems have explored multi-touch interfaces to control physical robot agents [14,15]. In our work we try to understand how simple and direct multi-touch interaction could allow the user to create motion paths that affected subsets of robots, which would in turn affect others. We present a multi-touch interface that gives users the flexibility to use natural gestures, such as those defined [17], to control a swarm of biologically inspired robots. We leverage research on biological motion [4], multi-robot pursuit evasion, and decentralized control [9].

3   Implementation

We implemented a multi-touch interface on an iPad app that supports up to ten fingers. The app is meant to control a swarm of semi-autonomous biologically inspired robots with simple behavior and low connectivity. Each robot can sense other robots within a certain range, and no communication between robots is assumed.

In our design three main components contribute to the swarm motion:

1. Biological data
2. User input
3. Local interaction rules.

When no input is given the swarm enters in a wandering status. While in this status the swarm follows a vector field derived from biological data stored in an xml file, i.e., the velocities of the individuals exactly match velocities measured from real swarms, flocks, etc.

When users touch the screen they start controlling the swarm. The fingers that touch the iPad create a virtual leader behavior at the position of the touch. Raising a finger from the surface corresponds to the removal of that leader from the swarm. As the user slides her fingers on the touch surface, the swarm reacts by following the leaders.

This behavior is coupled with additional interaction rules that make sure robots will avoid collisions and will try to stay with the group.

Following from [4], we define three different zones relative to each robot:

- Zone of repulsion (ZOR). Each individual attempts to maintain a minimum distance from others within a "zone of repulsion".
- Zone of orientation (ZOO). An individual will attempt to align itself with neighbors within the "zone of orientation".
- Zone of attraction (ZOA). An individual will attempt to move toward the position of leaders within the "zone of attraction".

The app provides a drop-down panel to set parameters and save them for future simulations (Fig. 1a). It is also possible to explore the effect of a parameter change in real-time, in effect, making the app usable for motion design.

We provide two different instances of the interface, one meant to test the evasion behavior and the other meant to test the pursuit behavior. In one instance, mobile pursuers try to hit the swarm and the goal is avoiding as much pursuers as possible. In the other instance, the iPad screen is filled with mobile targets, and participants are asked to control the swarm in order to catch as much targets as possible. Individuals in the swarm are represented as orange circles, while blue circles represents computer-controlled pursuers/evaders (Fig. 1). Below we give some examples of possible uses of this interface to design different kind of motions for the swarm.

One finger. One finger can be used to control the entire swarm (Fig. 1b). This strategy can be used to hide the swarm outside the field of view of pursuers (evasion task).
Two fingers. By using two fingers a line formation can be obtained (Fig. 1c). This formation can be used for pursuing. Another strategy is to split the main pack of robot in two subgroups (Fig. 1d). This strategy can be used both for evading and pursuing. When evading it is also possible to use a small number of individuals to draw the attention of the pursuers from the main group (Fig. 1e).
Three fingers. A V-formation, typically used for pursuing, can be obtained with three fingers (Fig. 1f). The same number of fingers can also be used to shape the swarm as a triangle (Fig. 1g).

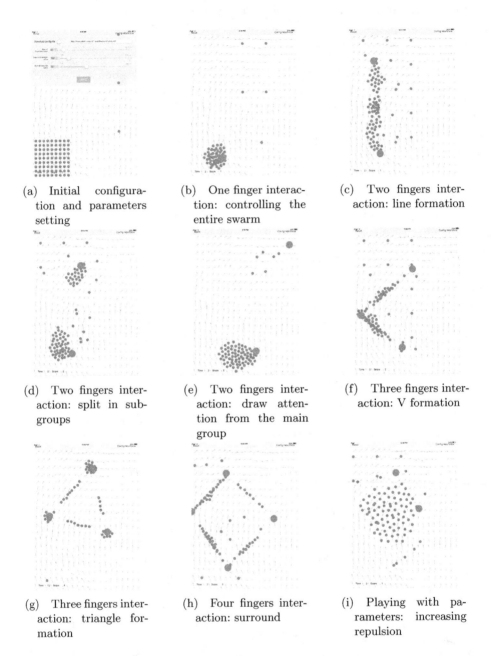

(a) Initial configuration and parameters setting

(b) One finger interaction: controlling the entire swarm

(c) Two fingers interaction: line formation

(d) Two fingers interaction: split in subgroups

(e) Two fingers interaction: draw attention from the main group

(f) Three fingers interaction: V formation

(g) Three fingers interaction: triangle formation

(h) Four fingers interaction: surround

(i) Playing with parameters: increasing repulsion

Fig. 1. iPad interface (Color figure online).

Four fingers. More than three fingers are typically necessary when trying to obtain complex configurations. Figure 1h shows how a surround behavior is obtained using four fingers.

Playing with parameters. A number of different kind of motion can also be obtained by playing with parameters. Figure 1h, for example shows the effects of an increment of the ZOR radius. This increases the repulsion component among robots of the swarm which causes a spreading behavior.

4   Demo Set Up

The setup comprises an iPad mini and a charger. In order to fully take advantage of the system, the user should be allowed to interact with the iPad in a comfortable way. A desk and a chair may be necessary to allow users to comfortably input commands even with two hands. Alternatively a flexible iPad stand to hold the device can be used.

5   Running Experiments

The platform described can be used to run different kind of experiments. In [1] it was used to understand how users would take advantage of the multi-finger control in the context of pursuit-evasion.

During the experiment, participants were given three to five minutes for each of the two tasks. Within that time, they performed multiple trials to iteratively refine the evading/pursuing strategy. Experiments containing iterative trials are useful to understand how people interact with swarms of autonomous robots. For example in [1], a similar setup was used to gain insights about frequently-used strategies for pursuit-evasion and the way they are achieved. Our work seeks to define new pursuit-evasion strategies and to suggest natural high level gestures that can be used to provide users with even more control over the interface while minimizing attentional demands.

References

1. Andolina, S., Forlizzi, J.: The design of interfaces for multi-robot path planning and control. In: 2014 IEEE Workshop on Advanced Robotics and its Social Impacts (ARSO), September 2014
2. Brooks, D., Yanco, H.: Design of a haptic joystick for shared robot control. In: 2012 7th ACM/IEEE International Conference on Human-Robot Interaction (HRI), pp. 113–114, March 2012
3. Cai, Z.-S., Sun, L.-N., Gao, H.-B., Zhou, P.-C., Piao, S.-H., Huang, Q.-C.: Multi-robot cooperative pursuit based on task bundle auctions. In: Xiong, C.-H., Liu, H., Huang, Y., Xiong, Y.L. (eds.) ICIRA 2008, Part I. LNCS (LNAI), vol. 5314, pp. 235–244. Springer, Heidelberg (2008)
4. Couzin, I.D., Krause, J., Franks, N.R., Levin, S.A.: Effective leadership and decision-making in animal groups on the move. Nature **433**(7025), 513–516 (2005)

5. Desaraju, V.R., How, J.P.: Decentralized path planning for multi-agent teams with complex constraints. Auton. Robots **32**(4), 385–403 (2012)
6. Drury, J., Scholtz, J., Yanco, H.: Awareness in human-robot interactions. In: 2003 IEEE International Conference on Systems, Man and Cybernetics, vol. 1, pp. 912–918, October 2003
7. Dunbar, W., Murray, R.: Model predictive control of coordinated multi-vehicle formations. In: 2002 Proceedings of the 41st IEEE Conference on Decision and Control, vol. 4, pp. 4631–4636, December 2002
8. Fong, T., Thorpe, C., Baur, C.: Multi-robot remote driving with collaborative control. IEEE Trans. Industr. Electron. **50**(4), 699–704 (2003)
9. Goodrich, M.A., Kerman, S., Pendleton, B., Sujit, P.B.: What types of interactions do bio-inspired robot swarms and flocks afford a human? In: Robotics: Science and Systems (2012)
10. Hayes, S., Hooten, E., Adams, J.: Multi-touch interaction for tasking robots. In: 2010 5th ACM/IEEE International Conference on Human-Robot Interaction (HRI), pp. 97–98, March 2010
11. Koenig, N., Howard, A.: Design and use paradigms for gazebo, an open-source multi-robot simulator. In: Proceedings of 2004 IEEE/RSJ International Conference on Intelligent Robots and Systems, (IROS 2004), vol. 3, pp. 2149–2154, September 2004
12. Likhachev, M., Stentz, A.: R* search. In: Proceedings of the 23rd National Conference on Artificial Intelligence, AAAI'08, vol. 1, pp. 344–350. AAAI Press (2008)
13. Flint, M., Polycarpou, M., Fernández-Gaucherand, E.: Cooperative path-planning for autonomous vehicles using dynamic programming. In: IFAC World Congress (2002)
14. Malik, S., Ranjan, A., Balakrishnan, R.: Interacting with large displays from a distance with vision-tracked multi-finger gestural input. In: Proceedings of the 18th Annual ACM Symposium on User Interface Software and Technology, UIST '05, pp. 43–52. ACM, New York (2005)
15. Micire, M., Drury, J.L., Keyes, B., Yanco, H.A.: Multi-touch interaction for robot control. In: Proceedings of the 14th International Conference on Intelligent User Interfaces, IUI '09, pp. 425–428. ACM, New York (2009)
16. Simov, B., Slutzki, G., LaValle, S.: Pursuit-evasion using beam detection. In: 2000 Proceedings of IEEE International Conference on Robotics and Automation, ICRA '00, vol. 2, pp. 1657–1662 (2000)
17. Wobbrock, J.O., Morris, M.R., Wilson, A.D.: User-defined gestures for surface computing. In: Proceedings of the SIGCHI Conference on Human Factors in Computing Systems, CHI '09, pp. 1083–1092. ACM, New York (2009)
18. Zhang, Y., Kuhn, L.D., Fromherz, M.P.J.: Improvements on ant routing for sensor networks. In: Dorigo, M., Birattari, M., Blum, C., Gambardella, L.M., Mondada, F., Stützle, T. (eds.) ANTS 2004. LNCS, vol. 3172, pp. 154–165. Springer, Heidelberg (2004)

Pointing and Selecting with Tactile Glove in 3D Environment

Yi-Ta Hsieh$^{(\boxtimes)}$, Antti Jylhä, and Giulio Jacucci

Gustaf Hällströmin katu 2b,
Helsinki, Finland
{yi-ta.hsieh,antti.jylha,giulio.jacucci}@helsinki.fi

Abstract. With emerging technologies, computing has evolved beyond spatial constrains, and location-aware applications have become possible. The symbiosis between human and computer has now a new factor: space. Aiming at increasing the awareness of the environment, we propose a tactile glove concept as a medium in between human and computer that allows computer to interpret users' intention in the space while information could be perceived by users via tactile modality. A glove prototype was built demonstrating pointing and selecting application in 3D space with tactile guidance. By following tactile guidance presented on the fingers, users can locate targets in 3D space without visual feedback from the system. As information presented on visual interface could dominate users' perception, which might hamper the awareness of the surrounding, the proposed glove interface implies an alternative that allows interacting with information while visual attention is released onto the environment.

Keywords: Tactile guidance · Symbiotic interaction · Motion sensing

1 Introduction

The future computing and symbiosis between human and computer envisioned by J.C.R. Licklider (1915–1990) more than 50 years ago [1] is finally emerging. Follow-up studies by Lesh et al. [2] also pointed out three elements that are required for symbiotic interaction: a complementary division of labor between human and computer, the representation in the computer of the user's abilities, intentions, and beliefs, and nonverbal communication modalities. Nowadays, computer is possible to exist not only in almost any form but also in any place. Location based applications and services have become available. By knowing the location of users, a computer has more contextual information and can provide more relevant information. The symbiotic relationship between human and computer has turned a new page when the interaction is no longer confined in a room.

To explore the interaction in the space under the context, we propose a tactile glove concept as a medium in between human and computer, which allows computer to interpret users' intention in the space while information could be perceived by users via tactile modality. Practical features of the glove would include tactile feedback as a presentation of information, and recognition of various hand gestures. Tactile feedback

© Springer International Publishing Switzerland 2014
G. Jacucci et al. (Eds.): Symbiotic 2014, LNCS 8820, pp. 133–137, 2014.
DOI: 10.1007/978-3-319-13500-7_12

has been utilized as a directional guidance cue in numerous studies, with devices ranging from handheld devices (e.g., [3–5]) to wearables such as belts (e.g., [6–8]) and gloves (e.g., [9–12]). In most studies, tactile cues for guidance have been designed with a specific application and use context in mind, thus, resulting in an increasing number of mappings and actuator setups.

In this demo, we present a tactile glove prototype and its use for pointing in target acquisition tasks in 3D space. Compared with previous work, our glove deploys wireless communication and does not require camera-based techniques for tracking hand movement, which provides high mobility and can be used in outside environment. We also implemented a demo application with a tourism scenario in which users use the glove to locate point of interests (POI) in the surrounding.

2   The Tactile Glove Prototype and System Design

We constructed a tactile glove for pointing in a 3D environment (see Fig. 1). Two Arduino microcontrollers (Arduino Pro) process the sensor signals and commands. For sensing hand orientation, the glove is equipped with a 9-axis Inertial Measurement Unit (IMU, InvenSense MPU-9150), which consists of a gyroscope, an accelerometer, and a compass. Flexible bend sensors (Spectra Symbol flex sensor) are deployed on three fingers (thumb, index, and middle finger). A set of four vibrotactile actuators (Precision Microdrives 10 mm shaftless vibration motor) is mounted for providing directional guidance. The actuators can be tuned to provide guidance in eight directions. The glove communicates wirelessly over Bluetooth with a laptop computer or an Android device running the application software.

The actuator placement is depicted in Fig. 2. Three actuators are mounted on three fingertips, and a fourth actuator is placed near the knuckle of the index finger. The vibration guides towards the direction where the user should point at. For example, when it vibrates on the thumb, it directs users to point toward left. When it vibrates on the index fingertip, the target is below the current pointing direction. When the user points at the target location, an on-target cue is presented as all four vibrators activated. Diagonal direction guidance can be achieved by actuating two vibrators at the same time. Detailed tactile guidance mapping is depicted in Fig. 3.

Studies have shown that differences in vibration burst interval are more perceivable than in intensity [6, 8]. We manipulated the burst interval to encode guidance cues. When the user was guided towards a target, vibration pulses with 100 ms activation and 400 ms silence were presented on the corresponding actuators, while pulses with 100 ms activation and 200 ms silence were presented on all actuators of a given set (fingertip or palm) when the user was on the correct target.

The system is designed to achieve high mobility to serve the purpose of pointing freely in 3D environment. The Bluetooth module on the glove functions as a serial communication port. Therefore, any devices with corresponding Bluetooth features could receive sensor readings from the glove and send commands to control the actuating of the vibrators. In our system setting, the glove is paired with an Android mobile device running our demo application.

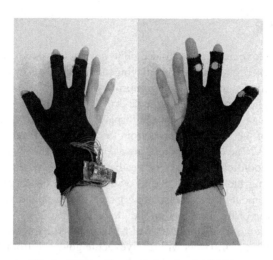

Fig. 1. The tactile glove showing both sides.

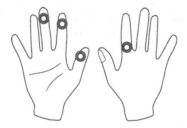

Fig. 2. Actuator placement on the hand. The colors depict ventral (blue) and dorsal (red) placement (Colour figure online).

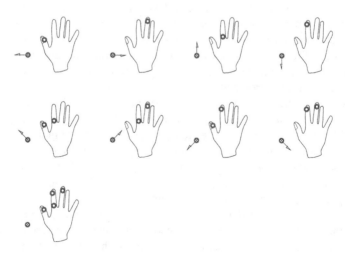

Fig. 3. Tactile guidance mappings on the finger. Colored circles indicate active vibrators (Colour figure online).

3   Demo Set Up

In addition to the glove and an Android mobile device, our demo setting also incorporates a loudspeaker or headset for playing audio content, which is added for enhancing the experience and expanding information capacity of our system.

An Android application running on the device is responsible for the computing and controls the behavior of the glove. On the display, the application shows in a radar-style visualization of the current readings from the glove. The hand movement can be observed accordingly. The target locations are also shown on the display. Each target is associated with a coordinate consists of heading and pitch reading from the glove sensor. By comparing the current pointing orientation and the target coordinate, the application could compute the corresponding tactile guidance and send commands to control the activation of the vibrators on the glove.

In our tourism demo scenario, users will be surrounded by several POIs. The first demo scene is locating POI. Users will receive tactile guidance from the glove, which directs the user to point at the target. When the pointing orientation matches the target coordinate, the on-target cue presents along with an audio annotation of the name of the POI is played. By bending the fingers, selection is confirmed and another audio recording with spoken information on the target POI is played back. The second demo scene is browsing POIs. Users receive no tactile guidance but can freely scan through the space. All the preset targets are available now. On-target cue is presented when pointing orientation matches any of the target coordinates.

4   Discussion

The glove could enhance users' awareness of the environment. Very often, our perception is dominated by information presented on visual interfaces, which might result in ignorance of the real environment. For example, it is common seeing people walking on the street while still using touch screen devices. Information is presented through visual elements and the interface also requires visual attention when operating the device. By using modalities other than vision, the awareness of the environment could be enhanced by direct interaction with real objects in the surrounding, e.g. point at objects and make selection gesture. In fact, visual information is actually presented by the real objects in the environment.

Moreover, the intention of glove users expressed through hand gestures and movement could be transmitted to computer for further interpretation. For example, pointing with hand and making a selection gesture could be interpreted as the will of fetching information from the pointed target. Equipped with Bluetooth modules, the glove is associated with computers or mobile devices wirelessly, which allows users to move freely in the space. High mobility presented on the glove could encourage users to more expressively interact with the environment.

Considering space as a resource in the symbiosis between human and computer, we argue that the glove concept has a role in cooperative use of the resource especially when combined with locating services. In the scenario where information is distributed in the space and in various locations, users' physical presence could contribute in gathering local data by leveraging mobility of the glove. Meanwhile, information could be fetched through natural interaction mediated by the glove.

5   Conclusion

In this paper, we present a tactile glove featuring sensing hand gestures and tactile feedback in pointing and selecting tasks. We also designed a demo with a tourism scenario in which our system assists users to locate POIs. Though visual feedback is not included in the system, the proposed glove has potential in increasing environmental awareness through the interaction and has a role in the symbiotic relationship between human and computer.

References

1. Licklider, J.C.R.: Man-computer symbiosis. IRE Trans. Hum. Factors Electron **HFE-1**(1), 4–11 (1960)
2. Lesh, N., Marks, J., Rich, C., Sidner C.L.: Man-computer symbiosis revisited: achieving natural communication and collaboration with computers. IEICE Trans. Inf. Syst. **E87-D**(6), 1290–1298 (2004)
3. Oron-Gilad, T., Downs, J.L., Gilson, R.D., Hancock, P.: Vibrotactile guidance cues for target acquisition. IEEE Trans. Syst. Man Cybern. Part C Appl. Rev. **37**(5), 993–1004 (2007)
4. Ahmaniemi, T.T., Lantz, V.T.: Augmented reality target finding based on tactile cues. In: Proceedings of the 2009 International Conference on Multimodal Interfaces, pp. 335–342. New York, NY, USA (2009)
5. Pielot, M., Poppinga, B., Boll, S.: PocketNavigator: vibro-tactile waypoint navigation for everyday mobile devices. In: Proceedings of the 12th International Conference on Human Computer Interaction with Mobile Devices and Services, pp. 423–426. New York, NY, USA (2010)
6. Srikulwong, M., O'Neill, E.: A comparative study of tactile representation techniques for landmarks on a wearable device. In: Proceedings of the SIGCHI Conference on Human Factors in Computing Systems, pp. 2029–2038. New York, NY, USA (2011)
7. Erp, J.B.F.V., Veen, H.A.H.C.V., Jansen, C., Dobbins, T.: Waypoint navigation with a vibrotactile waist belt. ACM Trans Appl Percept **2**(2), 106–117 (2005)
8. Asif, A., Heuten, W., Boll, S.: Exploring distance encodings with a tactile display to convey turn by turn information in automobiles. In: Proceedings of the 6th Nordic Conference on Human-Computer Interaction: Extending Boundaries, pp. 32–41. New York, NY, USA (2010)
9. Bial, D., Kern, D., Alt, F., Schmidt, A.: Enhancing outdoor navigation systems through vibrotactile feedback. In: CHI '11 Extended Abstracts on Human Factors in Computing Systems, pp. 1273–1278. New York, NY, USA (2011)
10. Gallotti, P., Raposo, A., Soares, L.: v-Glove: a 3D virtual touch interface. In: Proceedings of the 2011 XIII Symposium on Virtual Reality, pp. 242–251. Washington, DC, USA (2011)
11. Hein, A., Brell, M.: conTACT - A vibrotactile display for computer aided surgery. In: EuroHaptics Conference, 2007 and Symposium on Haptic Interfaces for Virtual Environment and Teleoperator Systems. World Haptics 2007. Second Joint, pp. 531–536. (2007)
12. Lehtinen, V., Oulasvirta, A., Salovaara, A., Nurmi, P.: Dynamic tactile guidance for visual search tasks. In: Proceedings of the 25th Annual ACM Symposium on User Interface Software and Technology, pp. 445–452. New York, NY, USA (2012)

Navigating Complex Information Spaces: A Portfolio Theory Approach

Payel Bandyopadhyay[1]([⊠]), Tuukka Ruotsalo[2], Antti Ukkonen[2], and Giulio Jacucci[1]

[1] Department of Computer Science, Helsinki Institute for Information Technology HIIT, University of Helsinki, Helsinki, Finland
{Payel.Bandyopadhyay,Giulio.Jacucci}@helsinki.fi
[2] Helsinki Institute for Information Technology HIIT, Aalto University, Espoo, Finland
{Tuukka.Ruotsalo,Antti.Ukkonen}@aalto.fi

Abstract. Users often find difficult to navigate through a large information space to find the required information. One of the reasons is the difficulty in designing systems that would present the user with an optimal set of navigation options to support varying information needs. As a solution, this paper proposes a method referred as *interaction portfolio theory*. This theory is inspired by the economic theory called the "Modern Portfolio theory", which offers users with optimal interaction options by taking into account user's goal expressed via interaction during the task, but also the risk related to a potentially suboptimal choice made by the user. The proposed method learns the relevant interaction options from user behavior interactively and optimizes relevance and diversity to allow the user to accomplish the task in a shorter interaction sequence. This theory can be applied to any IR system to help users to retrieve the required information easily.

Keywords: Information retrieval · Information retrieval systems · Modern portfolio theory · Search behaviour · Evaluation

1 Introduction

The stored information in the web is increasing rapidly. The main drawback of this is that users are often not satisfied with the results and face difficulty in finding the desired information [1–3]. There is a need to find novel approaches that better support the interaction between users and computers reducing uncertainty [4–6]. In IR system, like search engine, a typical interaction technique to express information needs is a *typed query* and then the system retrieve information from the information space based on that query. A common problem encountered by these systems is that users might use different terms for searching the same information which might lead to dis-satisfactory results. This problem is known as the *vocabulary mismatch problem* [7]. Often users are unable to form "good" queries especially in an information space unfamiliar to the user [8].

© Springer International Publishing Switzerland 2014
G. Jacucci et al. (Eds.): Symbiotic 2014, LNCS 8820, pp. 138–144, 2014.
DOI: 10.1007/978-3-319-13500-7_13

Another type of IR systems are where the system provides the user with relevant options and the user navigates the desired options to find the relevant information. Marchionini's [9] studies have shown that users prefer "recognition task" rather than "description task". In recognition task, information are organized in some order and the users navigate through the options to find the relevant information, often referred as "navigation" [10]. This approach is advantageous over the other as the user has a possibility of retrieving the document without specifying a "query" which is required to be matched with the given information space. These systems are suitable for casual users having no prior knowledge about the query [11].

In these type of systems, the user finds the required information by navigation rather than by searching. In *navigation* user is guided to explore the organization and contents of an information space. It is helpful for users, who are unable to provide the exact information they need, to explore the given information space to find the relevant information [12]. Some systems use hypertext to create a network of inter-related documents to help users to explore the information space [11]. Studies have shown that users often get disoriented in these systems [13]. In few systems, a hierarchical structure of information is presented to the user, through which user can navigate interactively to get the relevant information. Often this pre-defined organization is confusing for the user, as the term "well organised" is also ambiguous. The term "organization" varies from user to user: for a same set of information display, one user might find it easy to navigate through while it might be difficult for another user who have different information need and background [8].

To resolve the drawbacks of pre-organized information, this paper proposes a method to support navigation which is called *interaction portfolio theory*, inspired by an economic theory of financial investments called "Modern Portfolio theory" [14]. This method helps to adaptively select an optimal combination of interaction options for the user to accomplish a navigation task. The method involves a sequence of rounds, where N keywords are displayed in each round and the user clicks the relevant keyword to find the displayed target document from a complex information set. The main aim of the method is to provide the user with best combination of interaction options that eases the user to find the target document easily. Choosing the right combination of interaction options which is called as "optimum set" is difficult to find. The proposed method tries to solve the optimization problem of finding "pareto-optimal" set of interaction options where "relevance" and "diversity" are optimized at the same time to maximize gain of interaction options that can get user closer to her goal and minimize risk of choosing interaction options that are less relevant due to possibly suboptimal selections. A "pareto-optimal" interaction options set consists of a set of options, none of which are dominated by any other set of interaction options.

This paper is organized as follows. Section 2 provides an overview of the research work that has already been done in this area. Section 3 provides a more detailed overview of the proposed method for the formulated research question. Finally, Sect. 4 concludes the overall work described in this paper.

2   Background

To solve the problem of users getting lost in the vast information space (called disorientation problem), Hearst and Pedersen [15] have used clustering in their system called Scatter/Gather. The objective of clustering is to divide an unstructured set of objects into clusters (groups). This system clusters the whole information space to create categories which the user can select or gather. These categories are again merged together and the results are again clustered. The iterations continues on till the desired cluster of user information need is found.

Other solutions to the disorientation problem in navigation are content list [16], lattice representation [17], hierarchical representation [18] of information representation. Spatial maps provides the user a diagrammatic representation of the hypertext in a structure format. The diagrammatic representation acts as a feedback to the users having no knowledge about the information space. The drawback of this technique is the size of the information space. With the vast increasing data set in Internet, the pictorial view of the huge information space becomes difficult.

The main drawback of the above mentioned techniques is that these techniques of information retrieval are "static". Static in this context means that these techniques are pre-defined from before. Different users have different information needs and different background. Hence, a "well organised" hierarchical order of information might be very useful for an user to find the required information while another user might find it very difficult to find the information. These techniques suffers from the inability to adapt according to the user information needs [19].

There has been works related to "adaptive navigation support technology" [20] which creates user models based on user information need and eases the information searching. Joahims et al. [21] described a tour guide system which learns according to the user information need, Pazzani et al. [22] described a software agent that learns the user information needs, as user rates the relevance of information shown to the user. It changes its display results according to the ratings provided by the user. Kaplan et al. [23] have described a prototype which recommends user with relevant information according to the user preferences and needs of information. The results have shown that users were able to find the required information 40 % faster than using traditional IR techniques.

None of the techniques mentioned above have taken Modern Portfolio Theory (MPT) [14] into account to model the user interaction, so that a system can adapt accordingly. Wang and Zhu [24] have used portfolio theory in IR but the work is more concentrated in applying MPT in the ranking of "relevant documents". Traditionally in IR, only the relevance prediction of documents were used for ranking of documents. So, their work applied MPT in finding the "right combination of relevant documents" that should be displayed, so that the user finds the relevant document easily. The method proposed in this paper is more concentrated in finding the right combination of optimal interaction options which will help users to find the required information. This method can be applied to any IR systems using "recognition task" to retrieve documents.

The original theory by Harry Markowitz in his Nobel Prize winning work, stated that to maximize the expected return for a given amount of portfolio risk, one should carefully choose the proportions of various assets. The theory was based on the following two observations:

1. the future return of the invested stocks are uncertain
2. the future return of the invested stocks are correlated to each other

Taking the above two observations into consideration, the theory suggests that risk related to a certain portfolio investment can be nullified by using variance of the return. During investment, an investor should try to maximize the return and minimize the variance. The theory provides a solution to the problem that given a set of available securities or assets, which is the optimum way to invest money in these assets so that the investor gets the maximum return.

3   Proposed Method

In this paper, the main challenge lies in finding the optimal interaction options hence, the analogy of finance and IR is formulated as:

1. the interaction options which an user will find relevant to interact with the information space to find the required information are uncertain
2. the interaction options in the information space are correlated to each other

For example, to find a document related to "human computer interaction": users having a background of psychology might prefer a set of interaction options as relevant while computer science background users might prefer some different set of interaction options as relevant while users having some different background, other than those mentioned above, might find another set of interaction options as relevant to find the target document. Hence, though a document related to "human computer interaction" might have a particular set of options related to it, it is useful to find the optimal set of interaction options so that the results are useful to all users having any background. This optimal set of interaction options is formulated with help of the proposed method called as *interaction portfolio theory*, which is based on MPT. This paper describes portfolios as subsets of the set of all N set of interaction options, modeled as:

$$X = (X_1, \ldots, X_N), \tag{1}$$

where the binary variables X_i indicate whether interaction option set i should be included in the portfolio. A set of interaction option say X_i is chosen over other sets of interaction option when that interaction option set dominates other sets. If set X_i is not dominated by any other set, then this set is said to be non-dominated set. In multi-objective optimisation problem, there exists many such non-dominated set. These non-dominated optimal interaction sets on whole is called pareto-optimal solution. The proposed method focuses in determining the pareto-optimal set of interaction options where "relevance" and "diversity" are maximum. In this context, relevance means how related are the interaction options, given the user information need, while diversity is about the similarity of the interaction options to each other. The goal of the proposed method is to find an optimal set of interaction options that is relevant with respect to the information need but also heterogeneous enough to allow the user to pursue different directions, if needed.

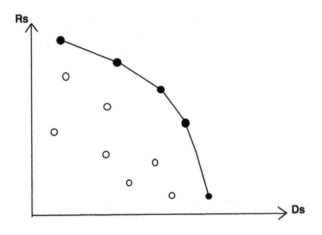

Fig. 1. An example of dominated and non-dominated interaction sets

Figure 1 demonstrates a set of dominated interaction option sets (portfolios) in black dots and a set of non-dominated interaction option sets (portfolios) in white dots. The dominance is based on two characteristics : Relevance (R_s) and Diversity (D_s). The black colored dots represents interaction option sets which are non-dominated and hence are called pareto-optimal solution sets.

4   Conclusion

This paper proposes a method called *interaction portfolio theory* which solves the problem of navigation faced by users in finding information in a complex information space. The proposed method is based on an theory of finance called Modern Portfolio Theory. The proposed methods displays the user with optimal interaction options. While choosing the optimal interaction options the method adapts according to the user's goal expressed via interaction during the task and also considers the risk related to a potentially suboptimal choice made by the user.

Acknowledgment. This work has been partially funded by the EU through the project MindSee (FP7 ICT; Grant Agreement # 611570).

References

1. Belkin, N.: Some(what) grand challenges for information retrieval. SIGIR Forum **42**(1), 47–54 (2008). http://doi.acm.org/10.1145/1394251.1394261
2. Athukorala, K., Oulasvirta, A., Gowacka, D., Vreeken, J., Jacucci, G.: Narrow or broad? estimating subjective specificity in exploratory search. In: Proceedings of CIKM 2014. ACM (2014)
3. Athukorala, K., Hoggan, E., Lehti, A., Ruotsalo, T., Jacucci, G.: Information seeking behaviors of computer scientists: challenges for electronic literature search tools. Proc. Am. Soc. Inf. Sci. Technol. **50**(1), 1–11 (2013)
4. Jacucci, G., Spagnolli, A., Freeman, J., Gamberini, L.: Symbiotic interaction: a critical definition and comparison to other human-computer paradigms. In: Jacucci, G., et al. (eds.) Symbiotic 2014. LNCS, vol. 8820, pp. 3–20. Springer, Heidelberg (2014)
5. Ruotsalo, T., Peltonen, J., Eugster, Manuel J.A. Gowacka, D, Konyushkova, K., Athukorala, K., Kosunen, I., Reijonen, A, Myllymki, P., Jacucci, G., Kaski, S.: Directing exploratory search with interactive intent modeling. In: ACM International Conference on Information and Knowledge Management (2013)
6. Glowacka, D., Ruotsalo, T., Konuyshkova, K., Kaski, S., Jacucci, G.: Directing exploratory search: reinforcement learning from user interactions with keywords. In: Proceedings of the 2013 International Conference on Intelligent User Interfaces, pp. 117–128. ACM (2013)
7. Deerwester, S., Dumais, S., Furnas, G., Landauer, T., Harshman, R.: Indexing by latent semantic analysis. J. Am. Soc. Inf. Sci. **41**(6), 391–407 (1990)
8. Bowman, M., Danzig, B., Manber, U., Schwartz, F.: Scalable Internet resource discovery: research problems and approaches. Commun. ACM-Assoc. Comput. Mach. - CACM **37**(8), 98–107 (1994)
9. Marchionini, G.: Interfaces for end-user information seeking. J. Am. Soc. Inf. Sci. **43**, 156–163 (1992)
10. Carpineto, C., Romano, G.: Information retrieval through hybrid navigation of lattice representations. Int. J. Hum.-Comput. Stud. **45**(5), 553–578 (1996). http://dx.doi.org/10.1006/ijhc.1996.0067
11. Marchionini, G., Shneiderman, B.: Finding facts vs. browsing, knowledge in hypertext systems. Computer **21**(1), 70–80 (1988). http://dx.doi.org/10.1109/2.222119
12. Burke, R., Hammond, K., Young, B.: Knowledge-based navigation of complex information spaces. In: Proceedings of the Thirteenth National Conference on Artificial Intelligence, vol. 1, pp. 462–468 (1996)
13. Elm, W., Woods, D.: Getting lost: a case study in interface design. In: Proceedings of the Human Factors Society 29th Annual Meeting, pp. 927–931 (1985)
14. Markowitz, H.: Portfolio selection. J. Finan. **7**(1), 77–91 (1952)
15. Hearst, M., Pedersen, J.: Reexamining the cluster hypothesis: scatter/gather on retrieval results. In: Proceedings of the 19th Annual International ACM SIGIR Conference on Research and Development in Information Retrieval, pp. 76–84. ACM, New York (1996). http://doi.acm.org/10.1145/243199.243216
16. Dee-Lucas, D., Larkin, L.: Learning from electronic texts: effects of interactive overviews for information access. Cogn. Instr. **13**(43), 431–468 (1995)
17. Carpineto, C., Romano, G.: A lattice conceptual clustering system and its application to browsing retrieval. Mach. Learn. **24**, 95–122 (1996)

18. Maarek, S., Berry, M., Kaiser, E.: An information retrieval approach for automatically constructing software libraries. IEEE Trans. Softw. Eng. **17**(8), 800–813 (1991). http://ieeexplore.ieee.org/stamp/stamp.jsp?tp=&arnumber=83915&isnumber=2739
19. Brusilovsky, P.: User modeling and user-adapted interaction. Adapt. Hypermedia **11**, 87–110 (2001). Kluwer Academic Publishers
20. Brusilovsky, P.: Adaptive educational systems on the World-Wide-Web: a review of available technologies. In: Proceedings of Workshop WWW-Based Tutoring at 4th International Conference on Intelligent Tutoring Systems (ITS'98), San Antonio, USA, 16–19 August 1998
21. Joachims, T., Freitag, D., Mitchell, T.: WebWatcher: a tour guide for the World Wide Web. In: Proceedings of 15th International Joint Conference on Artificial Intelligence, pp. 770–775 (1997)
22. Pazzani, M., Muramatsu, J., Billsus, D.: Syskill and webert: identifying interesting web sites. In: Proceedings of the Thirteen National Conference on Artificial Intelligence, Portland, pp. 54–61 (1996). http://www.ics.uci.edu/pazzani/Syskill.html
23. Kaplan, C., Fenwick, J., Chen, J.: Adaptive hypertext navigation based on user goals and context. User Model User-Adapt. Interact. **3**(3), 193–220 (1993)
24. Wang, J., Zhu, J.: Portfolio theory of information retrieval. In: Proceedings of the 32nd International ACM SIGIR Conference on Research and Development in Information Retrieval, pp. 115–122. ACM, New York (2009). http://doi.acm.org/10.1145/1571941.1571963

Author Index

Printed in the United States
By Bookmasters